EXCESS BAGGAGE

Leveling the Load and Changing the Workplace

Ellen Rosskam, Ph.D., MPH, BA
Visiting Professor, University of Massachusetts Lowell,
Work Environment Department
and
Visiting Senior Fellow, University of Surrey,
European Institute of Health & Medical Sciences, England

Critical Approaches in the Health Social Sciences Series
Series Editor: RAY H. ELLING

CRC Press
Taylor & Francis Group
Boca Raton London New York

CRC Press is an imprint of the
Taylor & Francis Group, an **informa** business

First published 2007 by International Labour Organization.

Published 2018 by CRC Press
Taylor & Francis Group
6000 Broken Sound Parkway NW, Suite 300
Boca Raton, FL 33487-2742

First issued in paperback 2018

ISBN 13: 978-0-415-78377-4 (pbk)
ISBN 13: 978-0-89503-360-4 (hbk)

Visit the Taylor & Francis Web site at
http://www.taylorandfrancis.com

and the CRC Press Web site at
http://www.crcpress.com

Library of Congress Catalog Number: 2006051663

Library of Congress Cataloging-in-Publication Data

Rosskam, Ellen, 1960-
    Excess baggage : leveling the load and changing the workplace / Ellen Rosskam.
        p. cm. -- (Critical approaches in the health social sciences)
    Includes bibliographical references and index.
    ISBN-13: 978-0-89503-360-4 (cloth)
    ISBN-10: 0-89503-360-7 (cloth)
  1. Airline check-in-agents--Health and hygiene--Canada. 2. Airline check-in agents--
Health and hygiene--Switzerland. I. Title.

    HD7269.A4262C27 2007
    363.11'93877364--dc22

                           2006051663

Dedication

For my father, Richard Henry Rosskam, the wisest man I've ever known, who survived all kinds of management practices, saw and heard it all (in spite of not), who carries little excess baggage, who taught me that what makes work so difficult is the people in it, and who always reminds me to wait and see how things come out in the wash.

For Alexander, for understanding the need and helping to slay the demons.

And to all those managers—for had you not tried so hard I might have stopped along the way.

Table of Contents

Foreword

Not all research projects, even those in occupational health, lead directly to concrete changes in working conditions. In fact, embarrassingly few do. And when women's occupational health is involved, research is rare and resulting change is even rarer. Why is it, then, that Dr. Rosskam's research, described in this book, has actually diminished the risk to check-in staff of suffering from musculoskeletal disorders? What is special about this research?

First, this is one of relatively few projects to focus on women workers. In 2000, Niedhammer et al. showed that only a minority of occupational health research projects involved women workers and even fewer analyzed the data on these women in any meaningful way [1]. In a recent update, we found that women's occupational health problems are still being neglected by researchers [2]. Those few studies that are done most often concern nurses and other health care workers. Much less is known about jobs like those at airport check-in, where employees deal directly with clients in other situations. Therefore, Rosskam's study is interesting to workers, unions, managers, scientists, and health professionals because it concerns people about whose jobs we know little, and rarely think about, even though we may see them very often.

Second, when the women report problems, Rosskam does not presume that they have the problems just because they are women. Instead, she looks for the source of the problems. It may seem normal for an occupational health researcher to think first about occupational causes for health problems, but in fact, research evoking vague biological or psychological causes for women's symptoms is all too common [2, 3]. Rosskam, however, takes care to uncover the factors in the women's working environment that are responsible for their health problems. Thus, she prepares the ground for prevention efforts.

Third, although the focus of this study is on musculoskeletal disorders (MSDs), Rosskam does not confine herself to one single source of suffering in the check-in workers's jobs. Instead, she discusses sources of stress holistically. "Stress" in the workplace is a complex construct, and the word itself can be used to refer both to a cause and its effect. Scientists are increasingly unpacking the concept of workplace stress, and they recognize that both women and men in the workplace suffer increasingly from health problems related to physical, emotional, and cognitive overload. They are also coming to realize that people are whole organisms, and their health is affected by all components of their surroundings, taken together [4, 5]. Musculoskeletal health is no exception, and Rosskam recognizes this in considering both physical and emotional stressors in this job. I am sure the check-in workers have recognized themselves in Rosskam's results, because the whole job and the whole person were considered. Why is this potentially important for promoting change? Because people are more likely to be mobilized for action by a study when they feel that the results are believable and valid.

Fourth, Rosskam also speaks directly to workers when she considers the strategies they use to keep from becoming ill from their work. She shows respect for them and involves them in the study. In her own words, she wants to identify "new ways of translating the concepts of worker empowerment into practice in today's work climate." She is interested in developing tools that will help workers to identify and resist excessive demands. This is an exciting new research subject that I hope she will pursue. The first results, reported here, show us some ways to combat workplace overload.

Fifth, Rosskam was able to obtain the collaboration of the unions in the workplaces that she studied. Unions have an interest in making change in the workplace, and that is their expertise as well. More and more often, unions and researchers in Europe and North America are coming to the realization that alliances between unions and researchers can help to identify occupational health risks that can be removed by changes in the workplace. The unions involved in Rosskam's study have been able to use the results to improve their members' jobs.

Sixth, Rosskam also collaborated with the employers, and her study speaks not only to workers, but also to managers, explaining to them the hidden costs of occupational health risks: in loss of worker competence, in sickness absence, and in worker dissatisfaction. Especially where service jobs are concerned, these costs are not obvious, since the worker is not manufacturing widgets whose rate of production can be monitored. Still, although an unhappy or ill service worker does not produce fewer widgets, she does become less able to meet the needs of her clients, especially when a reassuring smile is part of the clients' needs. Rosskam's study reminds managers of the important role of check-in workers in maintaining security and in calming the traveling public, and the need to care for the well-being of these employees.

Finally, unlike many other scientists, Rosskam does not mince words. A common conclusion even from researchers most devoted to improving workers' health is usually, "Further research is needed . . ." rather than "[Risk factor] should be removed immediately in order to alleviate the suffering. . . ." But

Rosskam's devotion to worker well-being leads her to a refreshing frankness that no doubt helped stimulate workplace participants to transform her conclusions into action.

We can see that Rosskam's book has led to change, but I am not sure that the jobs of check-in workers are even now entirely good for their health. More changes are surely needed, for the airport workers and for those in similar jobs. Most service workers still face long hours on their feet; unpleasant and even violent clients; poorly organized, repetitive work; and lack of respect from supervisors. Unfortunately, many occupational health scientists are still concentrating on the most visible, most dramatic workplace risks and neglecting the silent misery of many working women. I hope that this book will lead readers—scientists, managers, union activists, and workers—to take a new look at service workers and the challenges they face.

REFERENCES

1. I. Niedhammer, M. J. Saurel-Cubizolles, M. Piciotti, and S. Bonenfant, How is Sex Considered in Recent Epidemiological Publications on Occupational Risks? *Occupational and Environmental Medicine, 57,* pp. 521-527, 2000.
2. K. Messing and J. M. Stellman, Sex, Gender and Health: The Importance of Considering Mechanism, *Environmental Research.* Online publication: http://www.sciencedirect.com. Paper publication forthcoming. doi:10.1016/j.envres.2005.03.015.
3. C. Teiger, "Les femmes aussi ont un cerveau!" Le travail des femmes en ergonomie: Réflexions sur quelques paradoxes. *Travailler, 15,* p. 170, 2006.
4. S. L. Sauter and N. G. Swanson, An Ecological Model of Musculoskeletal Disorders in Office Workers, in *Beyond Biomechanics: Psychosocial Aspects of Musculoskeletal Disorders in Office Workers,* S. D. Moon and S. L. Sauter (eds.), Taylor and Francis Publishers, London, England, pp. 3-22, 1996.
5. S. Stock, N. Vézina, A. M. Seifert, and F. Tissot, Les troubles musculosquelettiques au Québec dans le contexte d'un monde du travail en mutation. *Santé, Société et Solidarité: Revue de l'observatoire franco-québécois de la santé et de la solidarité,* in press.

Acknowledgments

To Ray Elling for his enthusiasm, constant encouragement, and insightful comments on various drafts; to Karen Messing and Jeanne Stellman for their precious gift of time, valuable comments, and observations; to Pia Markkanen and Elisabeth Koestner for their much-appreciated intellectual input, belief in the need for this work, and for sustaining me throughout the process. My appreciation is due to Kishore; to Julie Lim for help with references and formatting; and to Bobbi, Julie, and all the other very kind and skilled people at Baywood for their assistance and guidance. I am deeply indebted to Sarah Wolfson for her excellent research assistance, and to Mary Lee Dunn who so skillfully shaped, pruned, and trimmed the fat.

My admiration is extended to Ricardo Semler, whose writing about it all helped me to stay on that road less traveled.

CHAPTER 1

The Insecurities of Service: Airport Check-In Workers

IMPACT OF GLOBALIZATION: SOMEONE ALWAYS PAYS

Airports around the world move millions of people and tons of luggage and commodities a day, a stream of motion from here to there, 'round-the-clock action that is powered by everyday men and women doing their jobs in an industry undergoing widespread and serious change. That power frequently is channeled through a strong back or a taut arm muscle. Sometimes it means sitting at an awkward console for hours endlessly checking in the public. Or having the wits to forestall a disgruntled would-be passenger's loud tirade over lost baggage or a changed travel schedule. While the jobs can be onerous, some workers find airport work exciting, even glamorous, serving as so many people's bridge to exotic destinations, important meetings, or homes somewhere else, rubbing shoulders with unnamed but important somebodies, the elite jet captains in their crisp uniforms, the sharp flight attendants with their caps cocked just so. It is enough to keep some workers in their airport jobs despite worsening conditions of work and pains in too many places.

Rapid change in an industry is tough on workers and working conditions. In a fierce globalized economy, airlines are experiencing increasing international competition, mergers, alliances, and cost-efficiency strategies that heighten the pressure on workers, managers, and companies. The pressures have direct and indirect impacts on jobs and the conditions of work in the sector. The social effects of liberalization and technological change in this complex industry are not making work any better for civil aviation staff. The companies see low growth while the workers experience general deterioration in the quality of existing jobs due to growing employment insecurity, atypical employment patterns, and longer work hours [1]. The introduction of electronic ticketing in the airports of industrialized countries, for example, is a change meant to virtually eliminate the need for check-in agents, other than the few required to help passengers who are unable to use the electronic apparatus. The industry's cost-cutting steps cause unemployment and weaken bargaining positions for unionized workers. In jobs that already have been de-professionalized over nearly twenty years through lowered status, pay, training and task delineation, such changes lead to work intensification, a further de-skilling of check-in workers, and eventually elimination of the workers' jobs. The low-wage practices of globalization may allow for the low ticket prices available today, but the low fares are not free. Someone always pays. While travelers pay with fewer comforts, less space, and declining service, the airline staffs bear the daily brunt of it. In the industry today, it is workers who pay the most when jobs are downgraded or cut, forcing the remaining workers to do two or three jobs. Management meanwhile demands that work tasks be done faster, with a smile and a cheerful relaxed tone of voice for all customers. High demands and deteriorating conditions produce physical and emotional strains on workers. The increased pace and intensity of work has been linked to rising numbers of work-related injuries and illnesses [2, 3].

LIBERALIZATION: ORGANIZING WORKERS INTO CONSTRUCTING THEIR OWN DEMISE

Airport check-in workers are not exempt from the reach of these forces. A check-in worker in California vented her frustrations over such pressures after helping a passenger at an electronic ticketing machine:

> This is how they get us to do more and more work faster and faster with fewer and fewer people. I hate it, but there's nothing I can do. They just put in these brand new machines. They expect that in the future everyone will know how to use them. Even the old people are expected to use them alone in the future. Today we still have to help a lot of people. That's why I'm here—to teach people how to use this machine that will end up replacing me. I hate it. (Worker, Los Angeles International Airport)

I was that Los Angeles traveler and I saw that check-in worker's anger, raw and very visible. Essentially, that worker had been organized into constructing her own demise.

In the mass media, airports have drawn significant, not always kind, attention since the September 11, 2001 attack on the United States. Measures taken in response to that catastrophe have imposed new burdens on the airport-as-workplace, burdens that have significantly expanded the pressures on airport work crews. The level of risk has gone up, for one thing, as shown when a sky marshal shot an airline passenger to death in Miami, Florida, in December 2005. The media have focused on airport security workers' and flight attendants' potentiality for preventing disruptive or dangerous passengers from boarding airplanes. Yet check-in workers could be a complementary front line to that of the security staff assigned to that task.

The complex workplace forces that cause ill health in airport check-in personnel are the focus of this book and the study that generated it. The study theorized that worker characteristics, physical and psycho-social factors in the work environment, and management practices have considerable impact on workers' health and well-being. The resultant physical and psychological stress influences the existence and degree of musculoskeletal injuries or disorders (MSDs) in the workers.

MULTI-FACTORIAL RESEARCH INVOLVING KEY "OTHERS"

This workplace study is multi-factorial. Our research team took a broad-based approach in designing the investigation, drawing in workers and their unions and employers at the three airports in Canada and Switzerland whose workers were studied, as well as key representatives on the national and international level. An important partner in the study was the tripartite Canadian Centre for Occupational Health and Safety (CCOHS); its constituents include trade unions, employers, and the national government. The work received support from the Canadian Auto Workers' Union (CAW), which represents check-in workers there, and the national trade unions PUSH and SSP/VPOD, which represent check-in workers in Switzerland. The Canadian Labor Congress and the International Transport Workers' Federation (ITF), with its 600,000 civil aviation workers in 110 countries, supported the research effort as well, as did the employers of all of the workers who were the subjects of the study. The inclusiveness of our approach that took both workers' and managers' viewpoints into the research facilitated the discovery of information and helped draw possible associations that otherwise would not have surfaced as part of the investigation. It enriched our understanding of the issues.

That inclusiveness also has implications for follow-up action on the study since it meant a greater level of "buy-in" from both sides of the bargaining table.

During many years spent traveling through airports in countries around the world, I had observed the work of check-in workers, how poorly designed their workstations often were, and that many lifted and carried a tremendous amount of heavy baggage. Discovering no significant body of research on check-in work, I approached the ITF to see if check-in workers appeared on their "list of priorities." Our discussions revealed that conducting research on this previously unstudied worker population was important for the International Transport Workers' Federation and its national level trade union affiliates. It was agreed that an exploratory study would be conducted in two countries, and the empirical evidence obtained would be used by collective bargaining agents of ITF affiliates worldwide. In the year that followed, discussions with unions at local, national, and international levels brought to the surface various issues, illuminating the intricacies of airport and check-in worker life. Discussions with management at the study sites generated rich information and interesting theories to explain the rising level of passenger-related violence on the ground and in the air, the complexities of managing an airport, and the increasing demands made by airlines.

PRACTICAL IMPLICATIONS FOR WORKERS AND EMPLOYERS

Since completion of the study, the results enable collective bargaining agents to negotiate for improved work conditions and ultimately better health for check-in workers with empirical evidence that previously did not exist. The study already has yielded direct practical implications for both workers and employers. The findings also enable employers at the study site airports and elsewhere to obtain a picture of check-in workers' work conditions, workers' health problems about which management may be unaware, workers' perceptions, and real and potential costs to employers associated with the conditions of work. From the design stage, employers at the study sites expressed interest in using the results for consideration of changes that may be needed.

The workers though were at the center of the research. The study's subjects were 132 check-in workers—a previously unstudied worker population—at three airports in two countries. They were service-sector workers, mostly, but not exclusively, women performing jobs that are high in demand but low in control, the kind of jobs which have been shown in other sectors to produce increased job strain that manifested as heart disease, anxiety, MSDs, depression, burnout, and other physical and psychological effects. (The Karasek and Theorell demand-control model demonstrates that jobs characterized by high psychological job demands combined with low decision-making autonomy over one's work are the most stressful types of jobs and lead to "job strain." Social support is a key resource for reducing stress-related disorders resulting from job strain, while low co-worker social support combined with high levels of demand and low worker control has been associated multiplicatively with cardiovascular disease prevalence [4, 5].) Check-in workers' jobs are low wage and low status; they involve fixed postures at non-adjustable work stations where workers can control neither the organization of work nor the design of the job, where both physical and psychological factors may aggravate or cause distress, strain and MSDs, where dealing with the public on a daily basis usually is an inherent part of what they do, and where exposure to violence is a high risk. And the workers often feel voiceless when it comes to industrial relations on the job.

The airports that participated in the study were chosen for their size, location, and the degree of mechanization of their baggage-handling systems. Semi-mechanized systems required the workers to lift and carry *every* bag checked in by passengers, whereas fully mechanized systems were meant to obviate the lifting and carrying of luggage. Those systems allowed a comparative analysis of the study's hypothesis, which asserted that more negative health outcomes occurred in check-in workers who used semi-mechanized baggage systems than showed up in those who worked with fully mechanized systems.

The study—which is explained in detail in later chapters—employed, briefly, participatory action research, a questionnaire, interviews, focus groups, work-station analysis, and statistical analysis. Its most significant findings were in two areas: first, the research revealed the prevalence and severity of musculo-skeletal disorders (MSDs) in the check-in workers, and second, the results disclosed the existence and degree of violence and the threat of violence to which check-in workers are exposed. Both findings break new ground in under-standing the burdens of these jobs.

MSDs are injuries to the muscles or skeletal system, frequently caused by poor work organization, awkward postures or movements and repetitive actions, such as assembly line work. MSDs generally are cumulative over time (frequency of exposure) and intensity of exposure. But not always. Discrete events, such as sudden and severe trauma from a single incident or movement, can produce MSDs even without long-term exposure. MSDs and repetitive strain injuries increased nearly 13% a year in the United States between 1982, when there were 3.6 per 10,000 workers, and 1999, when there were 27.3. By the turn of the century, MSDs had become the sources of the longest absences from work of all health and safety factors.

As suggested by the findings, the study did not confine itself only to physio-logical outcomes. Rather, the study team sought a holistic perspective on job factors that might cause ill health. Accordingly, as recognized today by progres-sive researchers and occupational health practitioners, our research looked for the physical and the psycho-social effects that might be due to both individual and organizational factors in the system. In application, this approach facilitated an investigation of jobs as entire work systems, in contrast with examining jobs exclusively by individual tasks. In focusing on the work system rather than the worker alone, the approach acknowledged the interactive nature of factors in the work environment.

A WORKER-BASED STUDY AT THE CROSSROADS OF NUMEROUS DISCIPLINES

Another departure from the usual in this kind of research was situating the study at the crossroads of several linked but different disciplines, including sociology, psychology, and occupational medicine. Applying a systems perspec-tive necessitated integrating aspects of all of these disciplines in order to gain new insight into the cause of MSDs in the worker population. Management issues also were integrated so as to examine an employer's perspective on improving work conditions. The approach flowed from the existing scientific literature, which demonstrated a need to explore the *multi-factorial* etiology of MSDs by focusing more on management practices and psycho-social factors in their development and severity.

Doing so was an important, even determinative, decision for the research. It built on previous studies which had identified five common risk factors for work-related MSDs based on exposure to them reported over the largest proportion of the working population:

1. High-demand job content;
2. Strenuous work pace;
3. Lack of influence and control over one's work;
4. Strenuous work postures and movements; and
5. Work-related violence (in service-sector jobs) [6].

Each of these risk factors is associated with unhealthy work conditions. In addition, for each of them, there are sociological, psychological, and medical aspects involved in how the facts surrounding the contributing issues should be analyzed. For example, mechanisms based on group dynamics and individual socio-cultural aspects affect how an individual approaches work or the degree to which one manages to live with pain from MSDs. Psychological factors determine coping mechanisms for exposure to ill-adapted and uncomfortable work conditions: distress and violence from aggressive passengers; disempowering management practices, such as those that suppress workers' voice, ignore worker empowerment options such as skills development and training, or quash workers' attempts to empower themselves by taking initiatives, using creativity, or attempting to take an active role in organizational decision-making. Medical aspects also must be considered in examining the physical aspects of work stations; work pace, tasks, and organization; and whether workers' health concerns are taken seriously or disregarded, directly or indirectly, by management. Such considerations are integrated into the examination of the variables in this study.

The investigation of all contributing factors increased the likelihood of more effectively discovering solutions to problems in any job with characteristics similar to check-in work and to improving work conditions and industrial relations in workplaces with similar characteristics. A reduction in costs to employers, workers, insurance companies, and society also may stem, directly or indirectly, from the outcomes.

The Smith and Sainfort (1989) balance model demonstrated the importance of taking into account physical, psycho-social, and organizational factors when examining worker health outcomes [7]. Further, their framework suggested that positive effects associated with the airport as a workplace might be strong enough to compensate for, or balance to some degree, the negative effects generated by other factors. Objective work conditions by themselves can contribute to adverse outcomes in worker populations. And particularly in the case of stress-induced health effects, a worker's *perception* that a job or task was monotonous, repetitive, or low in demand was associated with stress-related adverse health outcomes. In aiming to consider interventions for improving work conditions, both objective and subjective factors—that is, workers' own perceptions of their work—had to be examined.

By taking a systems approach, the study attempted to set itself apart from the large body of ergonomics-related workplace research to consider psycho-social factors and management practices together with physical factors as causes of MSDs. It was a distinguishing construct in our study. In earlier work, most ergonomics and work-related health researchers (particularly in North America) examined primarily physical factors alone or psycho-social factors by themselves. Few studies have looked at the impacts of both, combined with the effects of management practices, on worker health, as this one does. Management practices characterized as hierarchical and control-oriented that do not permit or promote the participation of workers on issues affecting their jobs are thought to contribute to MSDs. In fact, they are compounding factors, viewed in combination with biomechanical stressors, repetitive and undemanding work content, aggravating environmental factors, and psychological stressors, over long periods of exposure to such conditions. Increases in work intensity and quick turnaround policies (which dominate the civil aviation industry) appear to create similar pressures of work rhythm independently of the level of mechanization for checking in baggage.

If the occurrence and severity of MSDs was a major finding of the study, the second outcome dealt with workers' experience with and concern about violence or the threat of violence at work. Violence against check-in workers from aggressive passengers and managers is a growing phenomenon and a serious problem that has markedly increased in the past several years, including threats, verbal abuse and physical abuse. Both findings and recommendations for action are detailed in later chapters.

OVERCOMING OBSTACLES

The study grappled with several handicaps. In an innovation necessitated by the lack of scientific studies on the occupational health of check-in workers, surrogates were used to make comparisons, found in populations of mostly female supermarket check-out workers, computer clerical workers (again, primarily a female work force), and airport baggage handlers (usually male), and work tasks identified among each cohort that had commonalities with those of airport check-in workers. For instance, shared attributes might include, for the supermarket workers, shift work, dealing with the public, and little control over work pace. Both computer clerical workers and airport check-in staff worked at fixed, non-adjustable stations where they keyboarded into electronic equipment. And baggage handlers and check-in workers shared the same work environment, moved the same luggage, and did shift work. In the latter case, however, sex was an issue since baggage handlers are generally males, while check-in workers usually are females. It was not a small difference in those cohorts. For comparison's sake, identifying the surrogate worker populations was a productive, if not an exact, response to circumvent the problem.

The other significant obstacle of the study was the inadequacy of public health statistical information, which varied in important ways by airport and country, for this category of workers. Thus, the variations in several jurisdictions' record-keeping policies impeded comparisons or thwarted them altogether. There was no making up, however, for this deficiency in the archives of historical data.

Gender and sex were omni-present issues in the study [8], noted already in the comparison populations. The workers who thrilled to the excitement and glamour of airports—a positive psycho-social element for some workers—led the study to also consider gender-related influences on MSDs, detailed in chapter 3. The potential hazards of many jobs, particularly those performed by women, have not been examined by occupational health scientists because they are seen as "clean" or "safe" jobs or present situations that are difficult to quantify or otherwise describe. The dearth of a cumulative body of scientific knowledge on such jobs affects many women's positions in service-sector work. The pervasive and paternalistic inequality of sex roles has stereotyped women as working in what are viewed as "clean," less strenuous, often white-collar jobs, leaving the "dirty" hard labor in blue-collar work to men [9]. But appearances, as this study revealed, can be deceiving and, in fact, the reality of some women's "clean" jobs is camouflaged and actually is similar to that of industrial jobs performed by men.

APPLYING RESEARCH TO THE WORLD OF WORK: THE "BOTTOM LINE"

By testing such perceptions against fact, this study aimed to reduce the knowledge gap about the hazards associated with women's occupations, especially those whose characteristics are similar to check-in work. Indeed, our study's results contribute to the empirical evidence necessary to demonstrate that jobs widely viewed as safe and clean, and even glamorous, may not, however counterintuitive, be free of serious hazards. Studying check-in work can help us understand more about adverse health effects experienced by service-sector women workers everywhere that result from physical work factors, psycho-social factors, and management practices.

Gathering data and developing interpretations is one aspect of research, but applying the newfound knowledge is another, and one perhaps more meaningful to the research subjects. The study team wanted not only to generate new knowledge about the occupational health risks of check-in work but also to propose interventions that would improve workers' quality of health and life. The framework used in this study attempted to reveal where interventions might be most beneficial by, for instance, considering the actions and effectiveness of workplace health and safety committees. In later chapters, a number of suggestions are made for taking action to mitigate poor health outcomes in these jobs.

When it comes to acting, though, cost often sways decision-making. The price of taking action is sometimes the primary determinant in the decisions handed down from executive suites and board rooms where the bottom line is of paramount concern. The direct costs of worker health (or perhaps, more precisely, of "ill health") have always been germane in convincing companies of the importance of acting protectively. Indeed, many companies today measure such direct costs. The indirect costs of ill health, however, seldom are evaluated, usually are underestimated, and often are associated with the less well recognized causes of ill health, including management practices. Worker health has not been a priority concern of management in large part because companies have been required to pay only the costs related to illnesses that are recognized as work-related rather than to assume responsibility for the overall protection of health in the workplace. Yet a range of what are perhaps less well recognized physical and psycho-social workplace stressors, including exposures to various management practices and abusive managers, have been associated with ill health in workers. The costs of the range of physical and psycho-social effects that result are left to the worker and to society to bear; in essence, they are externalized from the workplace.

Since cost is such a controlling influence, adding it up is worth doing. Work-related MSDs are the most common women's occupational health problem [10] and today constitute the majority of occupational disease cases. Their high prevalence is compounded by the non-involvement of workers in the design stage of work spaces and equipment.

The consequences of the risks faced by check-in workers are costly for the workers, their companies, and society. Workplaces, work stations, and systems of work not designed from a human factors approach that involves workers from the design stage hand society at large, directly and indirectly, the burden of cost—more than $50 billion a year in North America and Europe. The European Union (EU) suggests that the total cost of MSDs to member taxpayers is billions of euros annually [11, 12]. Employers pay $15–$18 billion in workers' compensation costs annually [13]. Low back pain is especially expensive, accounting for 20–30% of all workers' compensation claims [14, 15]. In Japan, low back pain is one of the most common, costly, and serious occupational health problems, more than 60% of the number of occupational diseases that result from injuries [16].

Data from several countries illustrate the magnitude. The Nordic countries and The Netherlands estimate that work-related MSDs cost them 0.5% to 2% of GDP [17]. In Germany, work-induced MSDs account for nearly 30% of all work days lost to sickness; estimates say that lost work time costs 24 billion German Deutsche Marks a year. Britons lose nearly 10 million work days a year to work-related MSDs. Medical costs are estimated at 84 to 254 million pounds sterling, and direct and indirect costs to enterprises are estimated at 5,251 pounds sterling per injured worker (European Agency for Safety and Health at Work, 2000b) [17, 18]. In Canada, Ontario's workers' compensation board paid 171,047 workers in 1990 for lost work time due to injuries; half of them were due to work-related MSDs [19].

Costs are not low in the United States either. While conservative estimates vary, the government puts the costs of work-related MSDs as high as $50 billion annually and employers pay $15–$18 billion in workers' compensation costs every year [20]. The indirect costs from work-related MSDs include what we pay for lost productivity; uncompensated lost wages; personal losses, such as inability to function at normal physical levels and the related psychological impacts, marital break-ups, impacts on children and other family members, loss of homes, and so on; administration of programs such as household services for affected workers; lost tax revenues; and Social Security replacement benefits [20, p. 58). Industries that involve manual handling and repetitive work bear higher costs [21]. The combined direct and indirect costs of work-related MSDs were estimated at $149 billion for 1992 alone. The figures were considered *underestimates* of the burden [22, 23].

On a global level, the International Labor Office (ILO) reckoned that MSDs are 40% of the total costs worldwide of work-related injuries and diseases. It estimated that of the global burden of occupational injuries and diseases MSDs alone cost 1.6% of global gross domestic product [24].

WORKERS USED AS GUINEA PIGS IN REAL LIFE LABORATORIES

While it is well known that the health risks workers face vary by type of industry and type of job, considerable evidence shows that more flexible labor relations, and in particular the global shift toward external flexibility through downsizing, contracting out certain types of labor and so on, have been associated with deterioration in work conditions and workers' health.

A review of 190 studies from twenty-three countries (including developing countries, "transitional" economies, and industrialized countries) concluded that about 80% showed a link between decreased worker health protections and downsizing, greater employment insecurity, outsourcing, and the increased use of temporary and casual labor. More than 140 of the studies showed that precarious employment is associated with higher injury rates, hazardous exposures, disease, and work-related stress [25, 26]. Organizational downsizing, in particular, has been associated with MSDs [27].

Evidence showed that "new" management practices (such as "Quick Turn-around," which dominates the airline industry, "Lean Production," "Just-In-Time," "Total Quality Management," "Results-Based Management," and "Quality Circles") based on high work performance, work intensification, and psychological overdrive of workers, reduced workers' autonomy, workers' decision latitude, and individual control. These new systems of work organization have been introduced as a means of increasing productivity, quality, and profitability. Studies have shown that these "high performance" and "lean production" management practices have been associated with downsizing and restructuring [28]. Yet few studies have examined the impact of these management practices on occupational injuries and illnesses, or job strain, which has been linked to cardiovascular disease, hypertension, MSDs, and mental health disorders. The job strain literature provides a convincing body of evidence linking job strain with cardiovascular disease and hypertension in particular [29], but more study is necessary to examine health outcomes and "new" management practices, including outcomes due to abusive, dehumanizing, or disrespectful treatment by managers.

A review of fifteen studies on the impact of management practices dominated by work intensification found that across countries and industries psychological job demands had increased while worker control had decreased [30]. The emerging body of research provided growing evidence of high levels of perceived distress, MSDs, fatigue, and tension attributed to fast work pace, long hours, repetitive work, and few rest breaks. The result of these high strain management practices is that productivity gains come at the cost of worse health and diminished protection of the workers [27, 31, 32]. These management practices have been adopted in workplaces in countries around the world in the absence of a sufficient body of research demonstrating whether they are harmful in the long term. Essentially, workers have been, and still are, used as guinea pigs in a real-life laboratory where changes in the organization of work have outpaced awareness of their effects.

"MODERN" MANAGEMENT PRACTICES ARE MAKING WORKERS SICK

Research exposed the health risks of management practices based on work intensification, high strain, lack of worker control, and use of exclusively management control. For example, these "new" management practices resulted in much greater levels of MSDs and repetitive strain injuries caused by jobs that are high in demand with a low level of worker control; work under chronic stress; lack of worker involvement in any decision-making; poorly designed work tools, work stations, and work tasks; repetitive motion; static and/or awkward postures; and manipulation of heavy weights, to name a few key risk factors [33, 34].

Mental strain is related to time pressure and overwork. In its landmark mental health report in 2001, the World Health Organization (WHO) reiterated its prediction that there would be a dramatic rise in mental illness over the next twenty years; major depression already was the leading cause of disability around the world and was predicted to rise alarmingly over the next two decades [35]. Surely such findings are warning signals, indicating at minimum the need to pay more attention to jobs as part of an entire system. Equally worrisome, suicide is now seen as a major health problem, with a significantly higher rate among men in many countries. In Japan, there has been a significant spread of *karoshi*, or suicide from overwork. More than 30,000 people committed suicide in each of the last three years of the 20th century. Research conducted by Nishiyama and Johnson revealed that nearly half of their worker-subjects were anxious about their personal risk of karoshi [36].

Work intensification and overwork result in stress-induced adverse health outcomes, notably cardiovascular disease, mental health disorders, and MSDs. They have become the subject of trade union action and awareness-raising efforts [37, 38]. An independent magazine written for trade unions in the United Kingdom, collecting information from government sources and international unions revealed that work-induced heart attacks and death from overwork have been reported in numerous countries, including the United States, United Kingdom, India, New Zealand, Australia, China, the Philippines, Italy, Indonesia, Japan, and the Republic of Korea [39]. It is widely expected that the top work-related diseases of the 21st century will be heart attacks, suicide, and strokes. In examining the ever-evasive influences of genetics, lifestyle factors, and environmental exposures on various disease outcomes, the aforementioned work-related disorders and diseases can be understood to be largely socially constructed, not organically induced. (I find the following formula particularly useful: *genetics loads the gun while environment pulls the trigger.*) The distinction helps to re-situate the politics of occupational health and safety.

Lean production appears to have been singled out as a particularly insidious management practice in terms of worker health outcomes. Lean production has been assumed (by management) to be the solution to the industrial changes of the West. Lean production is claimed to offer challenging and rewarding work at the production level, fostering the re-involvement of rank-and-file workers. However, when the intensification of work under this system is examined, what has been claimed by management to be the introduction of "creative tension" has been called "mind-numbing stress" by workers [40]. The outcomes are hardly surprising when one considers management practices focused entirely on production, which discourage all forms of waste, including toilet breaks and time spent wiping sweat from one's brow [36, p. 630].

Corporations everywhere are motivated to undertake similar programs of quality management, waste elimination, and process improvement involving the workers themselves. However, most people who glorify [Japanese Production Management (JPM)] fail to consider that its focus is almost entirely on what benefits the company, not on what benefits the workers. JPM essentially increases management control and undermines the independence of labor unions and the human rights of the work force [36, p. 635].

Landsbergis, Cahill, and Schnall showed that after the introduction of full-speed production at one Japanese-owned car plant in the United States, injury and illness rates for 1988 were 66% higher than at similar plants, while the rate of cumulative trauma disorders was five times the industry average [30]. Workers in "lean" car factories reported significantly heavier workloads and too few workers compared with those in traditional companies. As well, low or decreasing decision-making authority was reported in "lean" companies. There was little evidence that workers in automobile manufacturing were empowered under lean production methods. In fact, the evidence was quite to the contrary—recent studies have confirmed that lean production intensifies the pace of work, while decision latitude remains low. Both factors contribute to job strain and job strain contributes to cardiovascular disease, hypertension, and MSDs. In jobs where biomechanical stressors are found, work intensification may lead to an increase in MSDs [41]; the exception found among those workers directly involved in a collective bargaining process [42].

I concur with Landsbergis et al. (1999) that any comparison of the costs and benefits of lean production and related work systems should incorporate the costs of chronic illness, including hypertension, cardiovascular disease, and MSDs. If illness is greater under lean production, or under any other form of production-intensified management practice, and if companies are held responsible, then there will be an economic incentive to moderate the risk factors for stress at work, such as emotional and physical demands, psychological harassment, and lack of decision-making autonomy. The same cost-benefit comparisons and moderations should be applied to all jobs, particularly emphasizing jobs that are high in demand with a low level of worker control due to their associated high level of job strain. Skorstad (1994) suggested that lean production has introduced a level of work intensification even *beyond* traditional Taylorist production methods [43].

A SCIENTIFIC CONSENSUS

Indeed, all that is new is not necessarily better. At least a consensus has emerged in the scientific world that a work environment characterized by authoritarian supervision, a delimited task structure, low work control with high demands, and repetitive and monotonous work may increase short-term productivity, but it has long-term adverse consequences for the work force.

Work has increasingly become subdivided into petty operations that do not sustain the interest or make use of the capacities of workers commensurate with their levels of education, knowledge, and experience. Airport check-in work, for example, is dissociated from the skills of the workers, the job subdivided into petty operations, and, according to the workers in this study, cannot sustain their interest. The work tasks are not designed to depend on the abilities of the workers, but rather on the management practices that determine entirely how the work is to be carried out. The long-term de-professionalization of the occupation has marginalized workers relatively as an occupational group and management has not recognized the value of the capacities, knowledge, and experience found among check-in workers. Quick turnaround practices have turned check-in work into a type of assembly-line factory job that obliges workers to check in passengers in increasingly short time frames—sometimes under three minutes per passenger. The human interaction element of the job is removed by the sheer force of time pressure, increasing work pace, work load, and tension, leading to psychological distress, MSDs, and high worker turnover.

The petty operations and tasks dominating "modern" ways of work demand less and less skill and training. The mindlessness built into modern work trends is alienating ever larger sections of the working population. Workers gain nothing from the fact that the decline in their control over their work process is compensated for by increased managerial control. Much service sector work has become more like factory work. Take supermarket check-out work as an example. Modern supermarket systems have greatly increased the number of customers that each worker can check out, and the systems are highly beneficial for inventory control, pricing, and so forth, but for the worker, the job is similar to factory assembly line-paced work, with external controls dominating tasks, tight supervision, repetitive movements due to the introduction of laser scanners, and intensified work pace limiting even toilet breaks. Training prerequisites for most service-sector jobs are minimal and workers have virtually no career opportunities. The jobs performed by computer clerical workers, supermarket check-out workers, and airport check-in workers are similar to those of continuous flow on a meatpacking assembly line or an automobile assembly conveyor. The piecework nature of such jobs makes control easy to exercise, using a variety of management methods, including electronic control.

The literature on technological change in modern work methods does not, in general, critique technological change as introducing new means of domination over working people, as a transfer from past forms of mechanized and bureaucratic domination to new forms of them [44]. It could be argued that the new forms are, in fact, more insidious when they are marketed and presented to workers as "team work," "empowerment-based training," "worker involvement," or "employee engagement," the latest management "buzzwords" in the United States In reality, they are merely euphemisms for new or additional forms of control over working people.

Findings from the mentioned studies pointed directly and indirectly to a growing variety of insecurities faced by workers, including employment insecurity, insecurity about representation of collective voice, income insecurity, skills development insecurity, and lack of worker health protections, including poor work conditions. Workers facing these various insecurities are less likely to be aware of the dangers of any type of work, less likely to be aware that they are exposed to any particular danger, less likely to be aware of needed preventive or corrective actions, less likely to be aware of the opportunities or means of taking appropriate action, less likely to be aware of the means of compensation, and less likely to be aware of the best means of coping and recuperating [45]. These facts make a strong argument in favor of holding companies responsible for the health and well-being of workers and for changing the current structural and cultural hegemony that produces work conditions that increasingly result in psychological distress and other adverse health outcomes in working people.

MARKETING "BAND-AID" APPROACHES

The massive development of what might be called "band-aids," i.e., palliatives to help working people deal with the adverse outcomes of the proliferating culprit of management practices, exists in part to perpetuate the growing insecurities. One of the greatest developers and promoters of worker-soothing "band-aids" is the pharmaceutical industry, which has a vested interest in maintaining (or even worsening) the status quo, since psychologically distressed, depressed, anxious, hurting, or ill workers purchase and consume medicines to get themselves through the day, to enable them to keep going to work no matter how ill they are, usually due to fear of job or income loss, or because they have no right to paid sick leave. In some circles, this is referred to as "presenteeism." One should not underestimate the importance of the linkage between social patterns of consumption and the dominant industry focus on "treatment" rather than prevention. Consider the example of occupational cancer. A phenomenal amount of money is poured into research on cancer treatment—far less than for research on the identification and elimination of carcinogens, that is, prevention. "Treatment" is an industry, generating profits for shareholders and CEOs (and jobs, of course). Quite simply, all-out prevention would necessarily alter things as they are and change usually is bound to upset someone.

One should also point to the "behavior-oriented" and "safety culture"—pushing safety "professionals" busily marketing "band-aid" approaches, such as organizational policies and training programs that are often thinly veiled individualized approaches to health and safety at work. Theirs is far from a *rights-based approach* to work and workers' health, which is essential to achieve decent and dignified work. The "safety culture" approach seldom addresses primary prevention (which requires tackling problems at the level of the organization). "Safety culture"-oriented themes and products often cleverly *appear* to promote prevention, while the focus is actually on changing workers' attitudes and behaviors. The approach "encourages" workers to engage in "partnerships" with employers to support voluntary initiatives, instead of regulation and enforcement. Resultant workplace policies, usually developed *without* the participation of workers, often focus on alcohol, drugs, and absenteeism, without addressing the *causes* of coping behaviors which frequently are rooted in management practices and organizational behavior [46]. Contributing to the structural and cultural hegemony, "safety culture" is being exported world wide, even by international organizations. Well-meaning, unsuspecting, governments, trade unions and NGOs in developing countries tend to jump on the "safety culture" bandwagon, unaware of the implications. But the bottom line is that "safety culture" has not delivered its promised results, a finding that deserves more attention and publicity. (Of course things can always be made to *appear* better if injury/illness reporting is non-existent or unreliable.)

Soon after the advent of computers in the workplace in the early 1970s, workers using them complained about routinized work, the drive for speed, and the unbearable pace [47]. Braverman cited the remarks of a computer clerical worker who was involved in a sociological study: "Although the girls do not quit, they stay home frequently and keep supplies of tranquilizers and aspirin at their desks" [47, p. 333]. Despite the costs and the damage, the risks have not been taken seriously. The "girls" and the aspirin were no match for the computer, which became ubiquitous in the workplace during the following years. So did certain workplace injuries.

ENDNOTES

1. C. Boyd and P. Bain, *A Summary of the Findings from the Cabin Crew Health and Working Environment Survey*, University of Strathclyde, United Kingdom in collaboration with Transport and General Workers Union and BASSA, 1999.
2. D. I. Levine, *Reinventing the Workplace*, Brookings Institute, Washington, DC, 1995.
3. P. S. Adler, B. Goldoftas, and D. I. Levine, Ergonomics, Employee Involvement and the Toyota Production System: A Case Study of NUMMI's Model Introduction, *Industrial and Labor Relations Review, 50*:3, pp. 416-437, 1997.
4. R. Karasek and T. Theorell, *Healthy Work: Stress, Productivity and the Reconstruction of Working Life*, Basic Books, New York, 1991.
5. R. Karasek, D. Baker, F. Marxer, A. Ahlbom, and T. Theorell, Job Decision Latitude, Job Demands and Cardiovascular Disease: A Prospective Study of Swedish Men, *American Journal of Public Health, 71*:7, pp. 694-705, 1981.
6. European Foundation for the Improvement of Living and Working Conditions, *Second European Survey of Working Conditions*, Luxembourg, 1996.
7. M. J. Smith and P. C. Sainfort, A Balance Theory of Job Design for Stress Reduction, *International Journal of Industrial Ergonomics, 4*, pp. 67-79, 1989.

8. 'Sex' refers to genetically determined differences while 'gender' is usually used to refer to socially determined male-female differences. Both sex and gender can have influences on manual handling capacity, although in manual handling, sex generally exercises a greater influence.

9. N. Sokoloff, *Between Money and Love: The Dialectics of Women's Home and Market Work*, Praeger, New York, p. 8, 1980.

10. National Institute for Occupational Safety and Health, Women's Safety and Health Issues at Work, NIOSH, U.S. Department of Health and Human Services, NIOSH Publication No. 123, 2001.

11. European Commission, *JANUS*, No. 20-11, 1995.

12. European Agency for Safety and Health at Work, *Research on Work-Related Low Back Disorders*, Office for Official Publications of the European Communities, Luxembourg, 2000a.

13. National Research Council, *Work-Related Musculoskeletal Disorders: Report, Workshop Summary, and Workshop Papers*, Washington DC, 1999.

14. D. M. Spengler, S. J. Bigos, N. A. Martin, J. Zeh, L. D. Fisher, and A. Nachemson, Back Injuries in Industry: A Retrospective Study. I. Overview and Cost Analysis, *Spine, 11*:3, pp. 241-245, 1986.

15. Statistics Canada, *Work Injuries 1992-1994*, Ministry of Industry, Ottawa, Canada, pp. 72-208, 1995.

16. S. Koda, S. Nakagiri, N. Yasuda, M. Toyota, and H. Ohara, Follow-Up Study of Preventive Effects on Low Back Pain at Worksites by Providing a Participatory Occupational Safety and Health Programme, *Industrial Health, 35*:2, 1997.

17. European Agency for Safety and Health at Work, *Inventory of Socio-Economic Information About Work-Related Musculoskeletal Disorders in the Member States of the European Union*, Belgium, October 2000b.

18. P. Buckle and J. Devereux, *Work-Related Neck and Upper Limb Musculoskeletal Disorders*, European Agency for Safety and Health at Work, Bilbao, 1999.

19. B. Choi, M. Levitsky, R. Lloyd, and I. Stones, Patterns and Risk Factors for Sprains and Strains in Ontario, Canada 1990: An Analysis of the Workplace Health and Safety Agency Database, *Journal of Occupational and Environmental Medicine, 38*:4, pp. 379-389, 1996

20. National Academy of Sciences, *Musculoskeletal Disorders and the Workplace: Low Back and Upper Extremities*, Washington, DC, p. 364, 2000.

21. B. Silverstein, E. Welp, N. Nelson, and J. Kalat, Claims Incidence of Work-Related Disorders of the Upper Extremities: Washington State, 1987 through 1995, *American Journal of Public Health, 88*:12, pp. 1827-1833, 1998.

22. J. P. Leigh, S. B. Markowitz, M. Fahs, C. Shin, and P. J. Landrigan, Occupational Injury and Illness in the United States. Estimates of Costs, Morbidity, and Mortality, *Archives of Internal Medicine, 157*:14, pp. 1559-1569, 1997.

23. Demonstrating the magnitude of the overall burden of occupational injuries and illnesses in the United States, data from 1992 reveal an estimated 13.2 million non-fatal work-related injuries. The direct costs of both fatal and non-fatal injuries were $49.17 billion; an additional $96.2 billion was estimated for indirect costs associated with wage losses, fringe benefits, lost production, recruiting and training replacement workers, rehabilitating affected workers, home production losses, and

time delays. (B. Silverstein, E. Welp, N. Nelson, and J. Kalat, Claims Incidence of Work-Related Disorders of the Upper Extremities: Washington State, 1987 through 1995, *American Journal of Public Health, 88*:12, pp. 1827-1833, 1998.)

24. International Labor Office, *InFocus Programme on Occupational Safety, Health and the Environment*, Geneva, 1999.

25. M. Quinlan, *Regulating Flexible Work and Organizational Arrangements*, paper presented at conference on Australian Occupational Health and Safety Regulation for the 21st Century, Gold Coast, July 2003.

26. M. Quinlan, C. Mayhew, and P. Bohle, The Global Expansion of Precarious Employment, Work Disorganisation, and Consequences for Occupational Health: A Review of Recent Research, *International Journal of Health Services, 31*:2, pp. 335-414, 2001.

27. P. A. Landsbergis, The Changing Organization of Work and the Safety and Health of Working People: A Commentary, *Journal of Occupational and Environmental Medicine, 45*:1, pp. 61-71, January 2003.

28. G. S. Lowe, G. Schellenberg, and H. S. Shannon, Correlates of Employees' Perceptions of a Healthy Work Environment, *American Journal of Health Promotion, 17*:6, pp. 390-399, 2003.

29. For examples of research showing these outcomes, see D. B. Baker, The Study of Stress at Work, *Annual Review of Public Health, 6*, pp. 367-381, 1985; J. E. Buring, D. A. Evans, M. Fiore, B. Rosner, and C. H. Hennekens, Occupations and Risks of Death from Coronary Heart Disease, *Journal of the American Medical Association, 258*, pp. 791-792, 1987; R. D. Caplan, S. Cobb, J. R. P. French Jr., R. Van Harrison, and S. R. Pinneau Jr., Job Demands and Worker Health, *National Institute for Occupational Safety and Health Publication No. 75*, p. 168, 1975; E. M. Cottington, K. A. Matthews, E. Talbott, and L. H. Kuller, Occupational Stress, Suppressed Anger, and Hypertension, *Psychosomatic Medicine,48*, pp. 249-260, 1986; S. G. Haynes, A. Z. LaCroix, and T. Lippin, The Effect of High Job Demands and Low Control on the Health of Employed Women, *Work Stress: Health Care Systems in the Workplace*, J. C. Quick (ed.), Praeger Scientific Publishers, New York; R. A. Karasek, Lower Health Risk with Increased Job Control Among White Collar Workers, *Journal of Organizational Behavior, 11*, pp. 171-185, 1990; R. A. Karasek, T. Theorell, J. E. Schwartz, P. L. Schnall, C. F. Pieper, and J. L. Michela, Job Characteristics in Relation to the Prevalence of Myocardial Infarction in the US Health Examination Survey (HES) and the Health and Nutrition Survey (HANES), *American Journal of Public Health, 78*, pp. 910-918, 1988; P. A. Landsbergis, P. L. Schnall, D. Dietz, R. Friedman, and T. G. Pickering, The Patterning of Psychological Attributes and Distress by Job Strain and Social Support in a Sample of Working Men, *Journal of Behavioral Medicine, 15*, pp. 379-405, 1992; P. A. Landsbergis, P. L. Schnall, J. E. Schwartz, K. Warren, and T. G. Pickering, Job Strain, Hypertension and Cardio-vascular Disease: A Review of the Empirical Evidence and Suggestions for Further Research, in *Job Stress 2000: Emerging Issues*, S. Sauter and G. P. Keita (eds.), American Psychological Association, Washington, DC, 2006; P. A. Landsbergis, S. Schurman, B. Israel, P. L.Schnall, M. Hugentobler, J. Cahill, and D. Baker, Job Stress and Heart Disease: Evidence and Strategies for Prevention, *New Solutions, 3*, pp. 42-58, 1993; S. L. Sauter, L. R. Muphy, and J. J. Hurrell, Prevention of Work

Related Psychological Disorders, *American Psychologist, 45,* pp. 1146-1158, 1990; J. Siegrist, R. Peter, A. Junge, P. Cremer, and D. Seidel, Low Status Control, High Effort at Work and Ischemic Heart Disease: Prospective Evidence from Blue-Collar Men, *Social Science and Medicine, 31,* pp. 1127-1134, 1990; R. Delbridge, P. Turnbull, and B. Wilkinson, Pushing Back the Frontiers: Management Control and Work Intensification under JIT/TQM Factory Regimes, *New Technology, Work and Employment,* 7:2, pp. 97–106, 1992; M. Parker and J. Slaughter, 1994; H. Bosma, M. G. Marmot, H. Hemingway, A. Nicholson, E. J. Brunner, and S. Stansfeld, Low Job Control and Risk of Coronary Heart Disease in the Whitehall II (prospective cohort), *British Medical Journal, 314,* pp. 558-565, 1997.

30. P. A. Landsbergis, J. Cahill, and P. Schnall, The Impact of Lean Production and Related New Systems of Work Organization on Worker Health, *Journal of Occupational Health Psychology,* 4:2, pp. 108-130, 1999.

31. See R. Delbridge, P. Turnbull, and B. Wilkinson, Pushing Back the Frontiers: Management Control and Work Intensification Under JIT/TQM Factory Regimes, *New Technology, Work and Employment,* 7:2, pp. 97-106, 1992.

32. M. Parker and J. Slaughter, *Working Smart: A Union Guide to Participation Programs and Reengineering,* Labor Notes, Detroit, 1994.

33. P. M. Bongers, R. R. de Winter, M. A. J. Kompier, and V. H. Hildebrandt, Psychosocial Factors at Work and Musculoskeletal Disease: A Review of the Literature, *Scandinavian Journal of Work, Environment & Health, 19:*5, pp. 297-312, 1993.

34. M. D. Brenner, D. Fairris, and J. Ruser, *Flexible Work Practices and Occupational Safety and Health: Exploring the Relationship Between Cumulative Trauma Disorders and Workplace Transformation,* University of Massachusetts, Amherst, January 2001.

35. World Health Organization, *The World Health Report 2001: Mental Health—New Understanding, New Hope,* Geneva, 2001.

36. K. Nishiyama and J. V. Johnson, Karoshi—Death from Overwork: Occupational Health Consequences of Japanese Production Management, *International Journal of Health Services, 27:*4, pp. 625-641, 1997.

37. *Stop Stress at Work,* Institute for Labor and the Community, New York, 1999.

38. *Enough Workplace Stress: Organizing for Change,* Canadian Union of Public Employees, Ottawa, 2003.

39. *Hazards Magazine,* No. 83, Sheffield, United Kingdom, July–Sept. 2003.

40. E. Skorstad, Lean Production, Conditions of Work and Worker Commitment, *Economic and Industrial Technology, 15,* pp. 429-455, 1994.

41. P. A. Landsbergis, P. L. Schnall, T. G. Pickering, K. Warren, and J. E. Schwartz, Life Course Exposure to Job Strain and Ambulatory Blood Pressure Among Men, *American Journal of Epidemiology, 157:*11, pp. 998-1006, 2003.

42. ". . . where biomechanical stressors are found, work intensification may lead to an increase in MSDs; the exception is those workers directly involved in a collective bargaining process" should be used with a caveat. The Landsbergis et al. (1999) study was based on a review of studies of new systems of work organization, such as lean production, which did not include ergonomics studies where biomechanical stressors and MSDs, within and not within the context of collective bargaining, may have come up. Notwithstanding, Kaminski, 1996 found *declines* in cumulative trauma disorders

among auto plant operating teams working under a collectively bargained contract, whereas other studies of lean production revealed *increases* in such disorders. The reduction in injury rates was attributed to job rotation, use of better ergonomic equipment, and a collectively bargained ergonomics program. (M. Kaminski, Wayne Integrated Stamping and Assembly Plant, Ford Motor Co./UAW Local 900, in *Making Change Happen: Six Cases of Unions and Companies Transforming their Workplaces,* in M. Kaminski, D. Bertell, M. Moye, and J. Yudken (eds.), Work and Technology Institute, Washington, D.C., pp. 25-44, 1996).

43. More discussion on this theme is found in chapter 6.
44. C. Korunka, A. Weiss, K-H. Huemer, and B. Karetta, The Effect of New Technologies on Job Satisfaction and Psychosomatic Complaints, *Applied Psychology: An International Review, 44*:2, pp. 123-142, 1995; A. Statham and E. Bravo, The Introduction of New Technology: Health Implications for Workers, *Women and Health, 16*:2, pp. 105-129, 1990.
45. *Economic Security for a Better World*, International Labor Office, Geneva, 2004.
46. P. L Schnall, P. A. Landsbergis, and D. Baker, Job Strain and Cardiovascular Disease, *Annual Review of Public Health, 15,* pp. 381-411, 1994.
47. H. Braverman, *Labor and Monopoly Capital: The Degradation of Work in the Twentieth Century*, Monthly Review Press, New York and London, p. 335, 1974.

CHAPTER 2

Check-In Workers:
A Forty-Second-Per-Passenger
Machine

The airport check-in work research team began with a virtual blank slate—the lack of scientific information in national databases and the lack of any significant work environment research record about this occupation. While the former imposed definite constraints, other scientists' lack of focus on airport check-in workers to date meant that we had both intellectual and methodological latitude in designing and orienting the study. We aimed to build an occupational health history for this set of workers. On that course, we found the voices of the workers.

The Annual Survey of Occupational Injuries and Illnesses conducted by the U.S. Bureau of Labor Statistics (BLS) is a useful national source of routinely collected information about occupational injuries and illnesses of workers. Occupational health statistics used or generated by the BLS may rely on or incorporate research conducted in other countries, which may provide the basis for inclusion of a particular occupational group. No studies or statistics on the occupational health of check-in workers were found through any BLS Annual Survey. Check-in workers, as an occupational group, do not appear in ISCO-88 (International Standard Classification of Occupations) [1].

The team's review of the scientific literature revealed very few investigations specific to airport check-in workers, which meant there was no reliable, clear picture that could be obtained of the extent of the problems faced by the workers. Among the few relevant studies identified, only one report—the 1996 Transportation and Communications Sector Review—addressed specific health and safety issues among check-in workers. The report said that too many check-in workers at a Canadian airport suffered from back injuries. The main cause identified was the baggage tag printer, which was located behind the counter so that workers had to twist their bodies to retrieve the tags. That repetitive motion caused workers to suffer back problems and a number of injuries were reported. The literature review also revealed that available statistical information is inadequate due to inconsistent frameworks for classifying injury and illness data. To establish the extent of occupational injuries and diseases and their associated costs among check-in workers, it would be necessary to approach all airlines individually, since they maintain their own statistical data.

REASONS FOR LACK OF PREVIOUS RESEARCH

Several explanations may account for the lack of prior research on check-in workers. For one, occupational health research and practice have been conditioned by the workers' compensation system. Occupational health and safety commissions define priority groups for study based on the degree of compensation to those groups. Research on workers' health tends to be concentrated on injuries and illnesses that cause lost work time and have a clearly defined cause [2]. This results in research generally focused on traditional male jobs in areas associated with high levels of workers' compensation. Consequently, many worker groups are excluded from such research. Workers who are not members of trade unions, for example, are not considered a priority group for research because they rarely make claims for compensation. As such, most workers' jobs in most sectors of industry are not studied for occupational health problems. This fact contrasts with the extent of research devoted to the workplace health of miners, construction workers, and chemical workers, all of whom have high rates of compensated occupational injuries and who, therefore, are frequent subjects of occupational health studies. Similarly, health problems that do not lead to claims for compensation are not considered research priorities and do not appear in occupational health statistics.

Worker turnover also inhibits investigation of many jobs, including check-in work. Many work-related health problems are cumulative in nature and reveal themselves only over time. Worker turnover can, in effect, "hide" problems and their original causes, making it difficult or impossible to relate present problems to past exposures. Job change may dissociate recent problems from past exposures, with no record at the original job of a problem that appeared later, but which actually is linked to that earlier work environment. The absence of records that show health problems linked to working conditions makes jobs *appear* "safe." Research does not become a priority for such jobs. These circumstances also make it difficult for workers to claim compensation for health problems that are work-induced. Workers may be left without recourse while employers are spared the burden of costs as well as their social and ethical responsibilities for protecting workers' health. What should be an additional concern to employers resulting from rapid turnover is that workers are more likely to have accidents at the start of a new job (often due to lack of familiarity with its requirements, lack of adequate training, and lack of vigilant supervision for new workers). Thus, higher turnover means increased accidents.

Most jobs performed by women are excluded from occupational health research. Airport check-in workers are mostly, but not exclusively, women. The "feminized" image of check-in work combined with the perception of check-in work as "safe" and "clean" help explain why check-in workers have not been included previously in work-related health or social and economic security studies. While dramatic, easily identified dangers are rare in the types of jobs usually assigned to women, nonetheless, many of these kinds of jobs present a multitude of work-related risks, including an important physical component that can produce pain and disability [3]. And indeed many jobs done by women raise multiple issues affecting their basic social and economic security. ". . . [T]he impact of waged work on millions of women in both the formal and informal sectors of the global economy continues to go unrecorded and unregulated" [4].

Research, recognition, and compensation are limited for women's work because the jobs usually assigned to women often lack substantial known dangers [3]. Men have from three to ten times more compensated industrial accidents and injuries per worker than women do, as workers' compensation statistics have demonstrated [5-8]. The statistics often are interpreted to mean that women's jobs are safer than men's, which may not be the case upon closer examination [3, 9].

Women often perform different jobs under different conditions than men, coupled with a heavier burden of domestic tasks, all of which must be considered when assessing the impact of waged work on women's health [10]. Indeed, women's waged work cannot be separated from the rest of their lives [11]. The male-female difference in biology is another factor to weigh. There are profound differences in the work capacities, particularly in the manual materials-handling capacities, of men and women, which support the consideration of both sex and gender as important factors in the etiology of MSDs, especially low back disorders [12].

Researchers should consider the effect of gender on how occupational health issues are experienced, expressed, defined, and addressed. Messing and colleagues detailed critical dimensions needed for gender-sensitive research [13]. They included the need for the sex of subjects to be reported in studies (in many occupational health publications, results are reported in such a way that the sex of the subjects cannot be ascertained), and for the interpretations of findings to acknowledge the limitations of the data regarding male-female differences, particularly since research conclusions have a role in shaping legislation and policy. But the most important difficulty is to reconcile the need for detailed data on exposure and outcomes. It requires large sample sizes to account for population diversity. Gaining access to workplaces large enough to make the distinctions often is difficult, which was a limitation of this study of check-in workers. Johnson and Hall echoed the need for work stress studies to analyze sex as well as class differences and noted that many such studies are restricted to males, while job characteristics found to be important stress predictors for males also are important for females [14].

In addition to methodological considerations, political and philosophical dimensions contribute to the lack of gender-based occupational health research. Research "purists" often argue that bringing up gender issues in research is "contamination," by introducing political issues into science. Further complicating the picture is the fear that identifying women-relevant occupational health issues will interfere with attempts for women to gain equality in the workplace. Notwithstanding, it is important to bear in mind the mission of occupational health research: *to prevent ill-health, disease and suffering among workers.* To this end, gender sensitivity is a means to increase the effectiveness of research. One cannot ignore the reality that occupational health is perhaps the most politically charged field of public health, imbued as it is with the politics of employers, trade unions, class struggles, structural hegemony, and calls for social change. Although gender-sensitive practices may be difficult to operationalize in some cases, they enrich the scientific quality of research and should generate better data and well targeted prevention programs.

PERCEPTIONS ASIDE, RISKS EXIST

While check-in workers have left virtually no track in the scientific literature, a number of obvious hazards are inherent in their work, as well as a variety of what may be less visibly apparent, but no less harmful, risks to the health of the workers. In health-related research, it is difficult to identify one factor that contributes to a health problem more than others, since numerous factors often help produce adverse health outcomes. Check-in workers face a clear set of risk factors for musculoskeletal injury, in particular, because they frequently lift and handle baggage and may operate a computer while standing or sitting in the same position for prolonged periods. MSDs are associated with physical risk factors, including the manner in which work is performed based on conceptualization of the job tasks; level of mechanization for baggage check-in; work content; physical postures adopted by workers as determined by job tasks and work station design; uneven workload distribution; inadequate or complete lack of training; the amount of time a worker is exposed to the conditions of the job; and exposure to various workplace environmental factors (such as extreme cold or heat, humidity, noise, insufficient lighting, or glare that inhibits viewing of the computer screen).

Other obvious hazards include exposure to particularly violent or aggressive passengers; a high level of job demand and a low level of worker decision-making latitude [15, 16]; uneven work scheduling; shift work and management-imposed shift assignment without worker involvement; and the emotional labor [17] resulting from interacting with the public when one is required to be polite and friendly at all times, including when faced with any kind of customer demand. Psychological distress results from violence; emotional demands; ill-adapted work station design and the related low level of comfort on the job, compounded by chronic pain from MSDs. It also results from the monotonous and repetitive nature of the job and from management practices based on control and work intensification, that exclude workers from decision-making, offer little decision-making latitude or autonomy, and allow workers to be treated without respect.

Still other risks affecting check-in workers include employment and income insecurity, the precarious nature of many jobs, lack of access to skills development, lack of consultation between workers and management, and little possibility to express one's voice as a check-in worker. Any or all of these factors may affect the existence and severity of MSDs and other ill-health outcomes.

While it is well known that the problems faced vary by type of industry and type of job, considerable evidence shows that more flexible labor relations and in particular the global shift towards external flexibility through downsizing, contracting out certain types of labor, and so on, have been associated with deterioration in work conditions and workers' health.

Time to Check Out Check-In Workers

This exploratory study theorized that groups of variables found in check-in work (worker characteristics, objective physical work environment factors, and management practices) have considerable impact on worker health and well-being. The resultant physical strain and psychological distress influences the existence and degree of MSDs, recognizing that these variables may individually or together cause MSDs. The fact that variables exert their effects in groups, rather than by single variable linear cause and effect, makes drawing links more difficult.

In order to look at the complex set of occupational health and safety issues among check-in workers in more than one country, as well as the costs of associated health problems, the study was designed using two countries to provide information for a comparative analysis. The research team sought employer injury records in order to identify lost work time rates due to MSDs so they could be compared among the three study site airports. A questionnaire was distributed to workers at all sites and both workers and supervisors were interviewed at each site in an attempt to identify any disparities between employer injury records and workers' self-reported, work-related injury experience. Discussions were held with groups of workers, trade unions and with management throughout the study process to identify possible solutions to problems. And surrogate populations of workers doing three similar jobs and who had been studied by other occupational health researchers were used for a comparative analysis [18].

The framework used in the study has relevance for policy-making. The study involved a variety of stakeholders in the research process (in support of the principle of participation of all involved parties). This broad-based approach was intended to catalyze a change process for the improvement of worker health and to improve management practices identified as having adverse impacts on worker health. Delegating responsibilities for worker health and safety, work design, and work organization to technical staff and technical systems alone is insufficient for monitoring conditions and making continuous or punctual workplace improvements.

This study contributes to the body of knowledge that calls for worker participation in designing work processes, worker contribution to decisions related to work organization, and job enrichment, particularly in jobs that are high-demand and low-control. Where management practices do not take these human needs into consideration, MSDs, as well as stress-induced adverse health outcomes, can be expected results. Job enrichment, encompassing skills development and training, such as preventive and empowering measures for dealing with aggressive or violent passengers, generates positive outcomes for workers and managers. Providing evidence from a previously unstudied worker population, this study aimed to contribute to knowledge about jobs with similar characteristics. Empirical evidence can help make convincing arguments that work should be more worker-centered and that management practices should be more human-centered.

Research on occupational health and safety from a management perspective is often limited to stress-related issues or legal issues—a limitation in the body of literature since the physical conditions of work receive little notice in the study of organizations. Risks to workers' health and well-being are associated with multifaceted, subjective interpretations, which may not be the same as those directly found in the work environment, many of which are linked to social inequalities [19]. The psycho-social links to both worker well-being and organizational outcomes deserve more attention. Research and theory suggest that individuals differ in their interpretations of risks, task characteristics, and work environments. This fact should be of interest to organizations, since worker morale, productivity, and turnover are affected [20]. Without exploring more deeply how workers' perceptions of health and safety issues in the workplace are influenced by individual attitudes and how they affect organizational outcomes, it is difficult to understand the intricate relationship between factors such as motivation and physical outcomes such as MSDs.

The approach applied in this study aimed to demonstrate the importance of, and impacts gained by, involving unions (and their health and safety committees) and management in action-oriented worker health investigations. Where health and safety committees are ineffective, it is useful to examine the plethora of reasons and obstacles that may account for it. Recognizing that there can be many reasons why a committee may not function effectively, and addressing them through workers, their unions, and management, can be a useful exercise. Peeling away the layers of information to identify causes and the related effects of ineffective health and safety mechanisms can catalyze action where blockages once dominated. The framework used in this study attempted to reveal where action might be most beneficial.

Objective work conditions are not alone in contributing to adverse outcomes in worker populations, particularly in the case of stress-induced health effects. Perception of a job or task as monotonous, repetitive, or low in demand was associated with psychological distress and adverse health outcomes. In aiming to consider interventions for improving work conditions, both objective and subjective factors (workers' perceptions of their work) must be examined.

Both research and management experience has demonstrated that a well-managed organization, which respects its workers, will be equally well-managed in its attention to the protection and promotion of workers' health. Chanlat has shown that organizations not well managed, not treating workers with respect, and not paying attention to work conditions, bear the costs of high rates of work-related accidents and illness [21]. Factors identified included the participation of workers, a system of work organization adapted to them, and a strategy of prevention integrated into the workplace and the organization as a whole.

Many managers prefer to ignore workers' health problems, including work-related stress, while there also are concerned managers who recognize the benefits of protecting workers' health. Pro-active managers actively seek out such information rather than passively wait for it. Eliminating the problem and at minimum addressing it openly can catalyze positive change for both workers and managers, leading to improvements in productivity and increased worker well-being. This kind of an approach is a key element of a pro-active management policy.

Conceptual Framework

Within the conceptual framework developed, check-in work is viewed as an entire system, where physical, organizational, psycho-social, and worker characteristics work conjointly as determining factors of MSDs.

While male and female check-in workers face exposure to the same work conditions, the significantly higher prevalence of women in the occupation dictates presumption of a strong gender dimension in airport check-in work. Characteristics that have been demonstrated to dominate in women's work, particularly in service-sector jobs, make check-in work a useful occupation to study to reduce the gap in knowledge about many jobs performed by women.

Little is known about women's jobs beyond the characterization of them as often high in demand, but lacking in worker control over the environment, organization, design of work, and how the job is performed [22]. Over the past couple of decades, there has been an important increase in the proportion of workers exposed to low decision-making autonomy in their jobs. Women, compared with men, are proportionally more likely to face this constraint, and among women this hazard is more often accompanied by a high level of psychological demand from work [23]. Airport check-in workers cannot define or determine how they do their work or the speed with which they must work. The manner in which they must handle baggage is conditioned by the system of mechanization existing at their airport site. The design of their work station is imposed, usually without consultation, and workers' feedback about their work stations is not requested. Management policies require polite and friendly treatment toward customers at all times, leaving workers few options.

An airport is an environment seldom studied, perhaps because working in that setting is seen as somewhat glamorous and work there is viewed as safe and clean. Yet, if an occupation has not been studied from a worker health perspective, such conclusions may be premature. In reality, the occupational hazards associated with service-sector jobs in airports are virtually unknown.

Airports are interesting environments in which to work (and to conduct research). Activity level is high, which creates a sense of movement and excitement. The face of the population that workers encounter changes daily, even hourly. Because the scenery changes so often, airports are not dull or boring places. Workers have space to walk around, often have a variety of shops and restaurants to frequent, which was the case in the three airports studied, and a large network of people working similar shifts with whom to interact. Workers experience a sense of great freedom walking around the airport and a feeling of satisfaction when each flight they have helped gets off the ground. There is still a certain glamour associated with flying, with aviation, long distance travel, airports, pilots, airplanes, flight crews, and airline uniforms. Airport check-in workers may be considered "status borrowers" to some extent by being associated with those who fly.

The attractiveness of the airport as a place of work, including the glamour associated with airports and with flying in general, may exercise a moderating or positive balancing effect on workers. The attractiveness of working in an airport may indeed diminish or dull the adverse effects of the negative factors, potentially decreasing the severity of resultant MSDs. Or it may simply explain why workers do not complain more frequently about MSDs.

Characteristics of being a woman worker in service-sector work include low wages, low levels of training, emotional labor, and lack of skills development or occupational development. Airport check-in work falls into this category, similar to computer clerical work and supermarket check-out work. The particular characteristics of being a woman worker are likely to contribute to the development or severity of MSDs.

Jobs characterized by management practices that impose a high level of demand on workers, while allowing workers little or no control over the various aspects of their jobs, and where there is a lack of consultation with workers, typify most jobs performed by women, and dominate service-sector and factory work. A lack of worker participation in organizational and workplace-based decision-making characterizes the management style that dominates check-in work as well as computer clerical work and supermarket check-out work. The presence of such management practices may contribute to MSDs, while the absence of these negative factors may be associated with fewer or less severe MSDs.

The combination of high level of demand on workers with a low level of decision-making autonomy in the job produces psychological distress and strain for workers. A higher level of psychological distress has been found among workers exposed to both high demand and low control, compared with workers not exposed to these two constraints [23, p. 578]. Lack of social support can leave workers in a vulnerable position and more prone to psychological disequilibrium. The types of jobs traditionally performed by women expose them, more often than men, to a lack of social support [24]. Among check-in workers, high demand is exerted by quick turnaround policies and the associated hegemonic trends of work intensification and increase in work speed through demands for rapid processing of passengers. Low level of control has numerous roots, including the lack of worker involvement in work design, work organization and organizational decision-making, and lack of consultation with workers on work station design. Equally disempowering is the lack of recognition of worker skill potential, lack of job and industry-related professional status, lack of career development opportunities, lack of training opportunities, and a labor market-generated preference for employing younger workers at low wages at the expense of senior-level workers with experience.

Risks posed to workers from passenger aggression and violence and lack of management policies and structures to address the problem of violence against check-in workers are also factors related to management style that are likely to contribute to MSDs. Finally, relations with management are included in the notion of management practices, which has a psycho-social component related to communication with workers. Where relations with management are not perceived by workers as supportive, management style may influence the existence of MSDs. In contrast, where communication channels are open, functioning *both* from top down and bottom up, and where management is perceived as supportive, management practices may diminish the prevalence of MSDs.

Action-Oriented Research Approach

Participatory action research (PAR) methodology was used to carry out this study, which was approached through a systems perspective. PAR methodology was selected to allow for participation by the unions and employers involved and because of the desire for the research to lead to positive change [25, 26]. An analysis of the study findings contributed to the body of knowledge on job-health associations and also generated a variety of proposed solutions to the problems identified in the study. The often-missing link between research and action in occupational health means that research findings seldom contribute to improving workers' health. Where such research is linked with trade union involvement though, there is a higher probability of sustained action to improve workers' well-being.

A consultation mechanism, therefore, was included in the study for conveying the study results back to both the unions and the employers at the study sites. Research based on PAR established a means of providing empirical data to collective bargaining agents, as well as giving employers of check-in workers a picture of the experiences of this group of workers. The link between research and action was built into this study at the design stage.

PROFILE OF THE METHODOLOGY

The research was conducted at three airports (2 in Canada—in Ontario and Québec, and 1 in Switzerland). Initially, observations were made about the conditions of work for check-in workers in various airports around the world. They were discussed with the ITF, followed by a preliminary literature review, which yielded no significant results pertaining to check-in workers. Discussions with the ITF (International Transport Workers' Federation) and its global affiliates highlighted a plethora of problems faced by check-in workers, related to a) conditions of work, with wide disparities among ITF's affiliates around the world, b) the "invisible" face of check-in workers in the civil aviation industry, performing what has been seen as a "glamour girl" job rather than viewed as a professional occupation, and c) management practices resulting in increasing marginalization of check-in workers.

The study included conducting a literature review, examining compensation records, surveying by questionnaire, interviews, work analysis, and focus group discussions with workers, employers, and airport authorities. The results of the study could help determine whether further research is justified, using a number of countries to give information for a multiple-country comparative analysis.

The selection of the three airports was based on their size and the system used for handling baggage to elicit a view of the range of baggage check-in systems, from manual to mechanized:

- Airport A was a small, regional airport in Ontario, Canada that had a semi-mechanized system for handling baggage. The baggage system required check-in workers to manually lift and carry every piece of checked baggage from the baggage scale up to the main conveyor, which then carried baggage farther along to the area where it was loaded into the airplane cargo hold;
- Airport B was a large international airport in Québec, Canada, that had a fully mechanized baggage system, where check-in workers could alternate work in a sitting or standing position. Its baggage system included a segment of conveyor connecting the baggage scale to the main conveyor, meant to eliminate the need for check-in workers to carry baggage manually;

- Airport C was a medium-sized international airport in Switzerland, with a fully mechanized system for handling baggage where check-in workers worked exclusively in a sitting position.

To study compensation records, published national statistics were examined and airport management authorities and airlines were asked to provide their own statistical data on injuries resulting in lost work time. To survey workers, a questionnaire was distributed to all check-in workers. Structured interviews with workers and supervisors were held to obtain detailed information. Work analysis was carried out by observing tasks and jobs, assessing work station layout, and videotaping workers as they performed their jobs.

Focus group discussions with unions at local, national, and international levels brought to the surface various issues, which helped to orient the study team to some of the intricacies of airport and check-in worker life that otherwise would be difficult for researchers to know. The study was designed in cooperation with the local, national and international unions and management. The local-level unions at the three study sites helped to make contact with the key management people in human resources. Following these links, the research team discussed and agreed with management and unions on the various steps of the study, how best to implement the questionnaire, conduct interviews, and create a feedback mechanism.

Management at the study sites provided other key contacts, such as individuals who could give information on the cost of installing and maintaining a fully mechanized baggage check-in system and on recorded work-related injuries and lost work time. At the initial stage of outlining the study design, the study team agreed with the ITF to examine a semi-mechanized baggage check-in system, a fully mechanized system where workers operated in a fixed position throughout their work shift, and a fully mechanized system where workers had postural flexibility. These criteria then were used to identify three airports of varying sizes so as to provide a broad picture of check-in conditions. The study team conducted its own outreach to obtain management permission and local union support to conduct the study at the various sites.

CARRYING OUT THE STUDY

The problems faced by check-in workers were first identified by the ITF but had not been quantified or studied in a comparative manner between airports. Since no previous studies had been conducted on check-in workers, the experience and knowledge of check-in workers and their representative unions and employers, including the Human Resource Departments, was relevant and guiding to the issues and the study overall. Input from all groups helped to shape the study design, while research objectivity was carefully maintained.

A workshop was held at the 1999 ITF international health and safety conference to discuss with global affiliates the issues they considered of priority importance for check-in workers. Participants mainly included ground staff from various countries. Working together, they defined the research questions and hypotheses. They adapted to check-in work the research instruments necessary to measure the various aspects and appropriate means of evaluating health and safety hazards and identified lost work time due to injuries and work-related illness.

Investigation was initiated with the support of management at the three study sites. Discussions with the employers supplied management perspectives on multiple issues, such as the various factors thought to contribute to violence toward check-in workers, management interpretation of worker-reported MSDs as compared with employer records on injuries and lost work time, mechanization of baggage check-in systems versus non-mechanization, training, skills development, empowerment of check-in workers, and systems of work organization. Similar discussions were held with the ITF, the national level trade unions, and local level unions, to obtain trade union perspectives on the issues. Management representatives expressed strong interest in using the study results to identify gaps in information and areas that needed improvement; unions expressed similar wishes, in addition to the ITF's stated intention to incorporate the study findings into its international media campaign against air rage for the protection of check-in workers against passenger violence. Preliminary testing of the questionnaire and interview schedule was performed at Airport A in Ontario, Canada.

The questionnaire was distributed across three airport sites and interviews were conducted with workers and supervisors at each airport. Data were analyzed with unions and management providing input on interpretation of the preliminary findings. The final phase of the study also involved preparing a report for both the unions and management and communicating the key findings back to them. A report for the ITF [27] was disseminated to the ITF's collective bargaining agents at its affiliates in 110 countries.

QUESTIONNAIRE

The questionnaire consisted of open- and closed-ended questions concerning work history (such as years in the profession), medical history including MSDs, physical work environment (for example, frequency of lifting), and psycho-social work environment (including violence at work).

The questionnaire was distributed to all check-in workers through management at the airports. Questionnaires were voluntarily and anonymously completed and returned by workers. A distribution cycle, calculated with management, was used to cover all shifts and capture as many workers as possible, including those cycling back to work after vacation periods. Protecting respondents' anonymity was important due to the personal nature of some of the questions, such as experience with violence, disclosure of MSD symptoms, severity of symptoms and whether they affected the ability to perform one's job, and the fact that management distributed the questionnaire. Management representatives at the central locations were engaged to remind workers to pick up and return the questionnaire, with the hope of increasing the response and highlighting management's support for the study.

Questionnaires were distributed in one round, to ensure confidentiality and protect workers' anonymity, a consequence of which was a lower response rate than might have been achieved with repeated surveying. The response rate was accepted in order to ensure anonymity and to maintain an un-pressured environment for workers' and managers' input to the study. In particular, it did not seem advisable to put pressure on management to engage in a second round of questionnaire distribution. Notwithstanding, the questionnaire distribution period was extended by two weeks at Airport C in Switzerland in an effort to increase the number of responses.

A body map was integrated into the questionnaire so that respondents could indicate on outlines of the front and back of a human body exactly where they experienced pain from MSDs, if pain was present. The body maps completed by the respondents helped to identify the extent of MSDs experienced by check-in workers [25].

QUESTIONNAIRE RESULTS

Questionnaires were voluntarily completed and returned by 132 workers:

- Airport A, Ontario, Canada: response rate 50.0% (8 workers of a total 16)
- Airport B, Québec, Canada: response rate 9.1% (32 workers of a total 350)
- Airport C, Switzerland: response rate 18.0% (92 workers of a total 504).

It is possible that the unions at each airport were not as rigorous as they might have been in encouraging workers to complete the questionnaire. Differences in response rates may be due to several other factors as well. Airport A in Ontario was used as the preliminary testing site for the questionnaire and interview schedule. Being a small airport, with only sixteen check-in staff, the study team was able to meet more of the workers personally during both preliminary testing and full launching of the questionnaire and interviews than at the larger airports. The response rate obtained may have been due in part to the particular work shifts that crossed with our presence.

The response rate from Airport C in Switzerland was better than that of Airport B in Québec due to a two-week extension of the response period, intended to increase responses. My own local presence may have had some degree of positive influence as well, proximity enabling me to speak numerous times with management, supervisors, and workers. Québec, where Airport B was located, was far from the Canadian research collaborators, who were based in Ontario. Travel to this study site was limited. Extending the response period at Airport B might have increased the response rate, however, requesting further intervention from management also might have been perceived as burdensome at the time.

The percentage for each question was calculated according to the number of answers for each particular question, rather than based on the total number of questionnaires received.

INTERVIEWS

Interviews with fifteen check-in workers (five at each airport, including at least one male at each site) and three supervisors (one at each airport) were guided with structured questions linked to the questionnaire. Workers were selected randomly, invited to a face-to-face interview, but decided independently whether to participate in interviews. Personal interviews allowed us to collect information about work-related health and safety aspects of the job, training, effectiveness of communication with employer, work scheduling, expectation of length of employment, and other concerns. Interviews were based on open-ended questions.

Overall, responses to interview questions were similar from all workers. Major concerns they reported included heavy workloads and poor environmental conditions. In general, responses indicated that a heavy workload is characteristic of the job, causing physical strain and psychological distress, with a very high volume of baggage to handle, with many peak and few slack periods, and where both passengers and management practices push the workers to work faster.

One management practice at the Swiss airport was directed at heavy workloads. Workers at this airport tried to follow the rule of having no more than one passenger in line for first class, three in line for business class, and seven in line for economy class. Auxiliary and part-time workers can be brought in during peak periods. Common complaints included stuffy air, uneven air temperature between check-in areas (some areas too cold, others not cool enough, some areas comfortable), poor lighting, and persistent noise from the baggage conveyor (Airports B and C). Addressing some basic health and safety issues, workers commented:

> The work station is not very safe, it has a lot of pointy corners. The first part of the belt is zig-zag, but the second part is smooth. The whole belt should be zig-zag. (Worker, Airport C)

> I am concerned about the cleanliness of the counters, the keyboard, and the passengers. Last year we requested vaccinations, but we were refused. (Worker, Airport B)

Highlighting health and safety concerns and lack of communication with management, one worker expressed this view:

> Management doesn't explain how health and safety is protected for the workers. I asked for information, but never got any answers. I have back problems. Suitcases fall down, even with the zig zag carpet because there is no bar. I have to pick up the bags and put them back on the conveyor belt. The chairs are not good, there's no full back support and they're not comfortable. The new chairs are not fully adjustable. The employer bought these without consulting some workers. On one side of the airport, the temperature is OK, but on the other side, nearer to the doors, it's too cold. There's a lot of noise, and it's tiring. The boarding pass printer makes noise and there is general noise all around. The lighting is very bright everywhere. The health and safety committee exists, but it's not that effective dealing with health and safety issues. At the beginning of this year, a worker was killed on the tarmac by a high-loader going backwards. It was noisy so the worker could not hear it coming. This should have gotten more attention than it did. It didn't change anything. (Worker, Airport C)

A general and consistent picture of check-in workers emerged from the face-to-face interviews conducted in Ontario, Québec, and Switzerland. In general, check-in work is the job of choice of both workers and supervisors. Check-in workers enjoy working with the public, are proud to wear the airline uniforms, and are committed to doing their best at all times. While perceptions about the effectiveness of communication with management differed among workers at the various airports, workers nonetheless expressed a general attitude of cooperation in the workplace.

Several "other concerns" mentioned by workers and supervisors included:

> Pushing wheelchairs with obese people through the snow, outside the airport in winter, is really hard. (Worker, Airport A)

One worker from Airport A and two from Airport B commented on the need to receive some training on handling passengers in wheelchairs.

One Airport C worker was concerned about having to work a shift beginning before 7 o'clock in the morning, obliging her to pay for a taxi to get to work since there was no bus at that hour. "If my schedule starts before 7 o'clock, I have to pay for a taxi. This is a problem because the taxi fare is really expensive for me," she said.

One Airport C worker summed up the major qualitative changes she noticed over time in her job and in the industry in general:

> I don't learn anything here. Only the computer system is used here, tag a bag, then explain things to people. Before, there were challenges, motivation, but not any longer. The ambiance changed. People who were here for 10-15

years are now leaving, and the pay cuts are demoralizing. The human touch is lost now. Making more money in less time is the only important point now. (Worker, Airport C)

Yet, in direct contrast with this comment, another worker at the Swiss airport said, "I find the job quite interesting. It's up to the individual to make the job interesting. If you don't like it, better leave it."

While the following health and safety-related comment may seem particular to Canada (or other countries where game hunting is common), it underscores the range of issues that check-in workers have to face:

I had a beef (no pun intended) with hunters bringing in their bloody trophies. I am a vegetarian. I wasn't sure if this was healthy and safe; for example, it made the floor slippery. I was told by management not to deal with those passengers and to have another agent help them. (Worker, Airport B)

These noted points highlight the need for worker participation in work organization and for effective use of workers' individual and collective "voice."

When asked about the effectiveness of the joint health and safety committee, both workers and supervisors at each airport made various comments:

I don't know if the health and safety committee is effective or not. (Supervisor, Airport C)

The health and safety committee is, for me, non-existent. It exists in fact, but it doesn't function. (Worker, Airport A)

I'm satisfied with the health and safety committee, but there's only so much they can do, due to infrastructure problems. (Worker, Airport B)

I know that the health and safety committee is trying to change some ways of doing our job. (Worker, Airport B)

I can be very involved in raising issues. I wrote a letter that the whole group signed, about the work stations. I did this because I had severe muscle pain from lifting bags. I wanted something done about the conveyor belt set up and the twisting we had to do. I discussed the problem with the supervisor and then got the petition letter going. I copied the letter to the health and safety committee with a couple of injury reports, but they never used it. I don't know who is the health and safety committee representative. (Worker, Airport A)

And two different workers at Airport B expressed concern about new scanners that had been introduced. Management had sent a letter introducing the new technology.

I'm concerned about the new scanners we got. We received a letter from management, but I don't believe them. I'm aware that this concern is with the health and safety committee. The cleanliness of the counters and keyboards is a problem. The health and safety committee was informed about this, but no action was taken. (Worker, Airport B)

I'm concerned about the new laser unit we got to scan boarding passes. The health and safety committee was approached about this, and a letter was sent out by management saying there was no danger, but some passengers told me the contrary. (Worker, Airport B)

Some comments during the interviews indicated mixed perceptions about communication with management. Here are examples from different individuals at each airport:

Pretty much there is open communication between management and workers. (Worker, Airport A)

I have almost never talked [about health and safety concerns] with my supervisor. I'm not very outgoing to report anything, and I'm not sure if there is any sense complaining. (Worker, Airport B)

There's no fixed communication, it comes up as necessary. Supervisors meet every week. (Supervisor, Airport C)

I think there's good communication between workers and management. (Worker, Airport C)

Communication between workers and the employer is up to the individual. You have to get a lot of information on your own. If a worker is quiet, you may not have good communication. (Worker, Airport C)

There has been a lot of staff turnover—about 25 people left recently. The employer wants to have people stay a maximum of two years in order not to pay higher wages. So senior people leave. They put pressure on senior staff to leave. Communication between workers and employer doesn't flow at all. The employer doesn't involve workers when communicating. My supervisor is very understanding. Not all supervisors are like this, not all will have the answers to your questions. You have to chase them usually to get your answers. (Worker, Airport C)

Workers at Airports A and B remain on the job longer, on average, than those at Airport C, or at least longer than those workers employed by one of the two employers at Airport C. When asked about their expectation for remaining on the job, responses from several workers at Airport A pointed to their long-term vision:

> I guess I will stay on this job until my retirement. (Worker, Airport A)

> This is pretty much a career job for me. (Worker, Airport A)

> I will probably do this job for twelve more years. (Worker, Airport A)

While work conditions appear worse at Airport A than at C, some factors appeared to override the negative conditions, motivating Airport A workers to stay up to ten years on the job, and still express enjoyment with their work. Notwithstanding, personal interviews revealed individual differences, demonstrated by the following comment made by one worker at Airport B:

> I guess I will do this job as long as I can. It's convenient for making some money. (Worker, Airport B)

Another worker at the same airport expressed similar sentiments.

Remarks made by five workers at Airport C revealed a lack of long-term expectation about remaining on the job. Some workers explained the reason(s) for their dissatisfaction:

> I've been doing this job for three years, and think I'll do this only until I find something else. I need more challenging work. This is a good job for part-time people and students. It's not a good job for full-time people. (Worker, Airport C)

> I've been here four years and will stay another two years. I was told that a normal passenger should take forty seconds to check-in. I used to be able to take more time with passengers. I want more contact time with passengers, to have them leave happy, not to be a forty-second-per-passenger machine. (Worker, Airport C)

> I've been on the job for one and a half years. I'm thinking about quitting now because the conditions are not good. I don't get the hours I want, the supervisor doesn't come to tell her issues to me, instead she goes to the unit head. They go behind my back instead of coming to me personally. (Worker, Airport C)

> When I was hired five years ago, I expected it to be for life. Now I think three to seven years is best. I might stay on past seven years, but I doubt it. (Worker, Airport C)

> I am leaving in one month. (Worker, Airport C)

Again highlighting individual differences, an Airport C worker expressed contentment with the job: "I've been working here for two years. No idea how long I'll be doing this job. It suits me well for now; it matches with my private life and gives me a lot of free time."

Perhaps, the most significant finding from interviews of both workers and supervisors was the unanimous confirmation of the lack of any ergonomics or safety training on either manual materials handling or computer use. Direct outcomes of the lack of basic knowledge about properly designed and adjusted work stations and of proper work techniques included:

- Work in awkward positions that exposed workers to discomfort with a higher risk for injuries;
- Limited improvements in work conditions as a consequence of lack of awareness of better ways of working.

Sample Population Demographics

Demographic factors (Figure 1) demonstrated that check-in work in Canada and Switzerland was predominantly, but not exclusively, a job performed by women. In 2000, of the 504 check-in workers at Airport C, 80% were women and 20% men. Of the sixteen check-in workers at Airport A, there were fifteen women and one man. Women constituted 72% of all 350 check-in workers at Airport B. Questionnaire responses were obtained from a 7:1 ratio of women to men at Airport A; a 28:4 ratio at Airport B; and an 80:12 ratio of women workers to

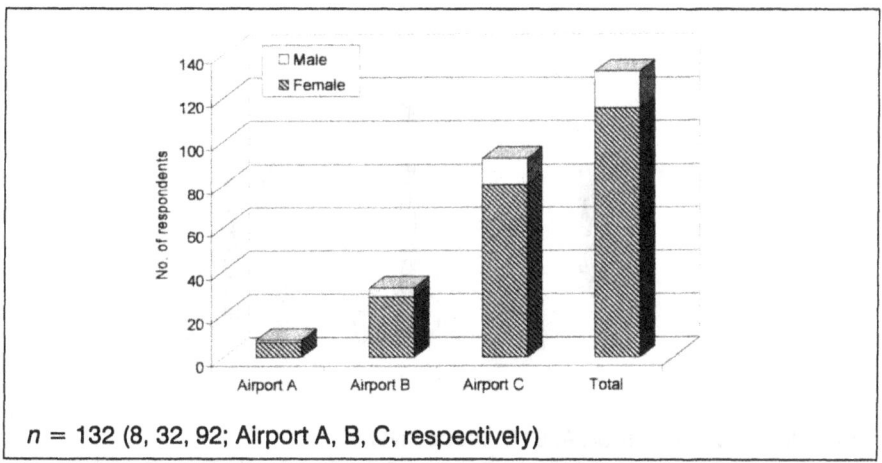

$n = 132$ (8, 32, 92; Airport A, B, C, respectively)

Figure 1. Demographic factors of respondents: Distribution by sex.

male workers at Airport C. Eighty-seven percent of the workers involved in the study were women, whose ages ranged from 31 to 46 years. Length of employment ranged from four to ten years on the job (Figure 2). The lower end of the range of employment duration is thought to be a result of one of the two employers at Airport C offering worse conditions of employment than the other, which logically might be accompanied by a higher degree of worker turnover.

Work Analysis

To understand the duties of check-in workers at all three study sites, the study team used direct observation of check-in workers performing their job functions and the positions they adopted while performing job tasks. Through observation, work analysis, and videotaping check-in workers as they worked, the principal components of the job were identified as: checking security and identification information, processing a passenger's ticket to generate a boarding pass, and expediting a passenger's baggage for loading onto the aircraft. No matter how similar these tasks might be at various airports, they can present very different workloads for the individuals performing them. Variations existed in the design of the computer work station and the baggage handling system, the volume of passengers, the quality of environmental conditions (lighting, noise, temperature) at any given airport, emotional demands from dealing with aggressive passengers (with additional variations between work at the first class counter compared with the economy class counter), number of agents available at the counters, language proficiency of check-in agents, and management practices.

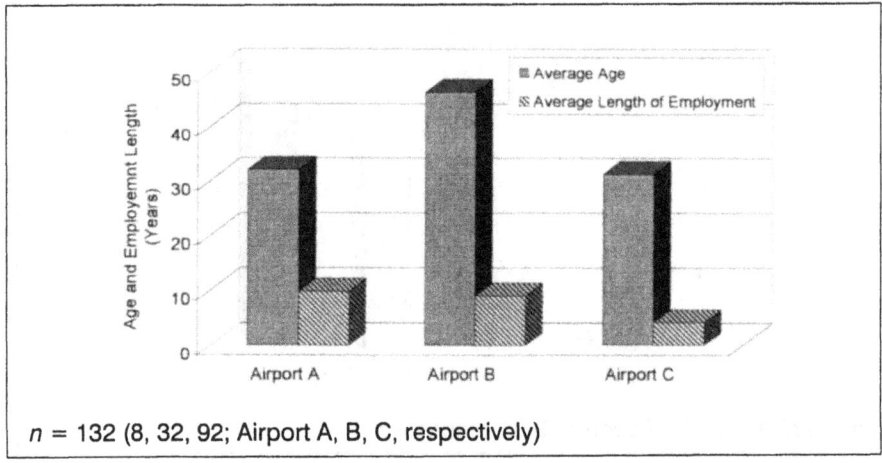

$n = 132$ (8, 32, 92; Airport A, B, C, respectively)

Figure 2. Average age and average length of employment.

COMPENSATION RECORDS

In order to examine published compensation records, national statistics in Canada were examined; the same attempt was made in Switzerland, but without much success. Airlines and airport management companies that consented to participate in the study were asked to provide their own statistical data about injuries resulting in lost work time. Since individual airlines were the employers of check-in workers at the two Canadian study sites, their consent was needed for participation in the study. At the Swiss site, management consent came through the two ground handling companies at Airport C, which employed the check-in workers.

In Canada, each jurisdiction had its own legal process for injury and disease claims. Between Canadian provinces, workers' compensation programs differ in benefits payable, administration practices, and in the compilation of statistics. There was no uniformity or consistency, since different workers' compensation boards used different classification methods based on their own needs. The study team obtained statistical data only about Canadian airport personnel in general, since no specific code existed for reporting lost work time injuries among check-in workers in Canada.

Compensation information about check-in workers in Switzerland was difficult to obtain for similar reasons, but there was one additional major circumstance—the unanticipated collapse of Swissair, which was the national airline of Switzerland at the time. Although Swissair was not the direct employer of check-in workers in Switzerland, it was the major employer of the ground handling company that employed the majority of check-in workers at Airport C and its demise affected collection of data for check-in workers in Switzerland. There, workers' compensation data are centralized with the Swiss National Insurance Fund (SUVA). Despite good will on the part of the airport ground handling companies, they were fully occupied due to the airline's financial crisis and not in a position to request this information from SUVA.

Adding to the difficulty of obtaining workers' compensation data for the two countries, neither Swiss nor Canadian workers' compensation data included a category for MSDs (the one exception was Québec, which has a diagnostic category for MSDs). This deficiency made it virtually impossible to identify work-related MSDs among *any* group of workers, even if the compensation data from the central insurance fund or compensation board were accessible.

Making Comparisons:
Airport Check-In Work and Three Similar Jobs

Studying jobs from a worker health and management perspective requires examining the management practices that are used in a given industry, or in direct relation to a particular job, and emphasizing their impacts on the health and well-being of workers in the particular industry or job. The lack of knowledge about the associations between jobs and workers' health stems mainly from the tendency of researchers to view jobs from a linear cause-and-effect perspective, looking at only one or two job factors rather than viewing jobs as part of a work *system*. Research approaches that examine single aspects of a job as the sole cause of negative health outcomes have not produced widespread improvement either in the way jobs are designed and performed or in workers' health. The gap in knowledge of associations between jobs and health also results from research approaches that seek only to generate knowledge but do not also envision intervention and improvement as direct outcomes.

No body of literature exists specifically related to airport check-in workers. Studies of airport workers' jobs focus on such tasks as loading and unloading aircraft, refueling, and de-icing, the effects of noise, chemicals, heavy physical work, the negative effects of night shifts, biomechanics for baggage handlers, and labor market information, including employment perspectives and requirements.

Because no previous research was undertaken on airport check-in workers, this investigation extrapolated from studies of surrogate populations, recognizing that similarities are useful only so far, after which there are points of departure. Working with surrogate populations offers value for purposes of comparison. It allows for the discovery of a foundation on which further knowledge can be built about different kinds of work experiences, for different groups of workers, and for the identification of potential solutions for problem-solving to improve the quality of work life. It encourages building on solutions that have been identified in previous studies. A review of studies of three comparison groups helps to reduce the knowledge gap about check-in workers, based on points of similarity.

The problems encountered by airport check-in workers are typical of other occupations with similar work-related health hazards, particularly those characterized by a high level of demand and low worker decision latitude. This cross-population review identified the major elements that have been studied in the comparison groups and has been weighed against the factors examined in the study of check-in workers.

Comparable Study Populations

Three comparison populations were selected based on similarity in work content, task, or conditions associated with the job. The selected populations included computer clerical workers, airport baggage handlers, and supermarket check-out workers; each group had job elements in common with check-in workers.

Computer clerical workers and supermarket check-out workers are typically female, low-wage workers, obliged to work in a fixed position either sitting or standing without the possibility of adjusting any element of the work station. The work is often marked by a high level of demand and low level of control.

Supermarket check-out workers have to deal with the public on a continuous flow basis.

Airport baggage handlers lift and carry baggage, similar to check-in workers at semi-mechanized check-in work stations. Baggage handlers are a predominantly male population, studied for levels of back injury in particular and lost work time due to work-related injury. The job has a high risk for physical strain.

Supermarket check-out work, computer clerical work, and airport check-in work are all low-wage jobs. All three place a high level of demand on workers, but with a low degree of control by workers in job design, organization, decision-making, and carrying out the job itself. These are job characteristics demonstrated to produce the highest level of strain and distress in workers [15, 28].

In some countries, such as Canada, supermarket check-out work is performed exclusively in a standing position. Check-in work, supermarket check-out work, and computer clerical work are all commonly associated with work stations, work tasks, and systems of work organization that are poorly designed or ill-adapted to a human being, with no possibility of adjusting any element of the work station, although sometimes adjustable chairs are provided. Both supermarket check-out work and computer clerical work are associated with high rates of MSDs and high rates of wrist injuries among computer clerical workers and supermarket check-out workers where laser scanners are used. MSDs among workers in these two jobs also result from the strain caused by high demand and low control. Additionally, both supermarket check-out work and computer clerical work are jobs that have a high degree of repetition and monotony and a high risk of psychological distress. All three of these service-sector jobs present additional risks and strain from regularly dealing with the public, that is, from selling one's "emotional labor" as part of the job [29].

Computer Clerical Workers

Research abounds on the impact of work with computers. Computer clerical workers have been more studied than the other two comparison groups due to the widespread introduction of computers over the last twenty years, marking a significant shift from Taylorism and Fordism to the technology-driven present era. Numerous studies have examined MSDs in computer workers from a bio-mechanical perspective and some studies have taken a multidisciplinary approach that considers a range of factors: stress and psycho-social factors, the impact of monotonous and repetitive work, lack of control over job activities, as well as the more obvious and measurable issues, such as visual strain and physical work station factors [30, 31].

A summary is provided of what is known about computer clerical workers, broken down into areas of similarity with airport check-in workers.

Since the early 1980s, research has shown that computer work is associated with a number of preventable risks [32, 33]. Research on MSDs related to work with visual display units (computer clerical) has generally focused on office workers, then from an ergonomic point of view led to an expanded vision of examining other jobs from similar points of view. The debate on the relationship between MSDs and computer-based work typically centers around occupational factors (constrained posture, poor physical workplace design, equipment design), work-related psycho-social factors (perceived high job demands, mundane, boring, repetitive job activity, little control, lack of support from colleagues and superiors), and bio-demographic characteristics (age, previous musculoskeletal injuries, emotional distress, family burden, and environmental factors).

Many of the negative effects associated with these groups of key factors are shown to be alleviated through worker and managerial participation in organizational decision-making, organizational change, and participatory problem-solving, with numerous benefits documented by key researchers in the field [34]. Yet despite a substantial body of information, there is insufficient knowledge based on studies using a multidisciplinary approach [34-36]. Standard research tools applied in a multidisciplinary approach included self-administered questionnaires (with questions specific to hand, wrist, neck, shoulder, and arm) [37], interviews, and standard ergonomic analysis of work stations and work tasks. The contribution of workers' ideas for work station improvements highlights the value of worker participation in the design process [34, 38].

Both computer work and airport check-in work that cause MSDs in the back, shoulder, arm and hand, are high risks for anxiety and psychological distress [39-41], and include characteristics typically associated with women's work [42-44], as well as work environment factors, stress-related factors, work design and work process factors, and job strain factors particular to service-sector work. Studies on computer workers concluded that work-related MSDs are a complex phenomenon that cannot be attributed solely to computer use but are related to the entire work environment.

More hours per day of computer use and less decision latitude on the job have been shown to be significant risk factors for potential MSDs. Support and conflict that workers with symptoms perceive in supervisor and co-worker relationships have been identified as potential risk factors [36]. High job demand, work organization issues, monotony and the repetitive nature of jobs have been identified as factors needing examination together with the work environment [34-36, 45-47]. Repetitive work often is thought of as only mentally fatiguing and spirit-breaking, yet this view fails to recognize the physically repetitive mechanism also involved in such work.

Lack of control over how their jobs are designed, carried out, and over the physical aspects of their jobs may be the critical factors directly causing negative reactions in workers facing job strain. These various mechanisms underlying the relationship between lack of control and MSDs constitute potent risk factors and deserve much more attention by both management and researchers. In relation to workplace stress, worker input to decision-making is known to be a key factor affecting how workers will react to workplace stressors, particularly to the introduction of new technology [34]. When new technology is introduced into the workplace, the degree of worker input in the planning process significantly affects worker satisfaction [34]. Evidence demonstrates beneficial results to both management and workers from workers' participation in the introduction of new technology.

Most of the conclusive research to date has examined more than one variable in examining adverse health outcomes related to computer work. A multi-variable approach is challenging, but useful given the many voids in understanding multifactorial problems for which no single solution fits all situations. The range of critical variables that should be examined includes:

- Work habits
- Task design
- Equipment and work station ergonomics
- Personal factors (e.g., sex and age)
- Individual psycho-social factors
- Organizational psycho-social factors, and
- Cultural psycho-social factors.

Few studies, however, attempt to examine the full range. Bettendorf et al., has been a strong proponent of the multi-variable approach to multi-factorial problems, demonstrating that studies that examine only one variable related to MSDs and computer work should *not* be considered conclusive [34]. The search for linear relationships between a single factor and an outcome leads to temporary solutions at best.

Discovering comparisons through what is known about computer workers, and approaches where questions of a multi-factorial nature are *not* examined, emphasizes the importance of applying a multidisciplinary approach, of looking at jobs from a systems perspective in work-related research. For example, examining only the physiological or psychological aspects linked to airport check-in work would not provide a well-rounded view of all work-related factors that may contribute to MSDs or reduce MSDs. Much of the research carried out to date lacks attention to organizational aspects of the work, while looking at worker-centered factors. This is a limitation in many study designs.

To illustrate this point, an approach that focuses strictly on biomechanical aspects of the job would not necessarily help us to understand the impact of aggressive passengers on check-in workers, or the impact of management practices that serve to marginalize workers, "encouraging" senior-level workers to leave, providing workers no skills development, no recognition for their work, and leaving their ego needs unfulfilled. Studies applying a strict biomechanical focus on computer workers have not examined the potential impact of psycho-social factors and work organization [48-51]. Research taking a multidisciplinary approach (such as combining a biomechanical focus with a psycho-analytical approach, e.g., [17, 34, 35, 52, 53]) helps in understanding much more about the variety of factors working together to cause adverse health outcomes in computer workers.

Studies finding that worker and managerial participation in organizational decision-making and organizational change alleviate the negative effects of job factors may come close to the findings of Smith and Sainfort's balance model of job stress [54]. The model provides a basis for reducing stress and its associated negative health consequences by "balancing" the various elements of the work system. The negative effects of physical aspects of a given job may diminish when those negative aspects are balanced with worker participation, including in solving work process, design, and work organization problems.

> . . . the adverse influences of low job content can be balanced by an organizational supervisory structure that promotes employee involvement and control over tasks. Proper job design can be achieved by providing those characteristics of each element that can meet recognized criteria for worker ego needs fulfillment and that set proper physiological and psychological loads . . . using elements to compensate for poor aspects in other elements can balance the stress with moderating factors that reduce the load and potential health consequences [54, p. 76].

Perhaps what Smith and Sainfort call "ego needs fulfillment" carries primary importance for addressing MSDs and work-related stress particularly in jobs characterized as high in demand and low in control and should be emphasized much more in organizational change strategies, worker health studies, and work-related studies of a multidisciplinary nature.

Supermarket Check-Out Workers

Following is a summary of what is known about supermarket check-out workers, broken down into areas of similarity with airport check-in workers.

Researchers started to investigate supermarket check-out work when electronic scanning became a suspected cause of carpel tunnel syndrome in check-out workers due to the repetitive wrist movement involved. Most of the research that has been conducted on supermarket check-out work is biomechanics-oriented, focusing on MSDs and quantitatively measurable physical work environment factors [55, 56]. The majority of studies substantiate the problem of MSDs among check-out workers and some describe interventions [57]. Most researchers have looked at worker-centered factors, but few have paid attention to organizational aspects of the work. Some of the main ergonomics-oriented studies include [58-62]. Studies taking a more multidimensional approach combined with intervention, some involving both workers and management, some recognizing psychological distress and psycho-social factors related to MSDs, and some examining issues of worker control over job activities, include Vézina et al. [63], Niedhammer et al. [64], Orgel et al. [65], and Toon [66]. In the range of studies evaluated, videotaping was useful for detailed examination of job activities.

Supermarket check-out work is a job with a number of similarities to airport check-in work. Both involve a high degree of static posture. Both jobs are usually performed at work stations that often are ill-conceived for human movement and comfort. Supermarket check-out workers in most countries of Europe, Latin America, Africa, and Asia work sitting down during an entire work shift, while Canadian and many U.S. check-out workers work only standing up [63]. Airport check-in workers work mostly either standing up during an entire work shift or sitting throughout the whole shift. The possibility to alternate between sitting and standing was found at only one of the study sites, and has been seldom observed in airports around the world. In general, supermarket check-out workers are given few breaks. Vézina and colleagues addressed symptoms and work conditions among check-out workers in Canada [63]. Of twenty-three workers observed, sixteen reported pain while working, particularly lower back and limb problems that were found to be related to prolonged standing. Our findings among check-in workers were similar. Supermarket check-out work is performed today mostly using an electronic scanner which registers both the price and stock code of items as they are purchased. Some workers still use conventional cash registers. But work stations are seldom ergonomically designed and chairs often lack adjustability, similar to the situation found among airport check-in workers.

Supermarket check-out work is the fourth most common job among women in Canada, and workers typically stand for long periods handling items at a fast pace. Characteristic of women's service-sector jobs, supermarket check-out workers are often assigned to fast-paced tasks involving repetitive movements and awkward postures, although the physical burden on these workers might appear slight and be classified as *light* work [63]. These are additional characteristics in common with airport check-in work, which also is seen as *light* work. No recognition is given to the thousands of kilograms a day that workers might lift and carry.

Inherent in supermarket check-out work as in airport check-in work, is the pressure to respond to the needs of customers. The emotional demands of being friendly, obliged to treat every customer politely and wear a pleasant face at all times, is an additional job stressor. While this psycho-social factor is associated with the job and risks causing a negative impact on workers' well-being [67], it has been little examined in studies of supermarket check-out workers. Soares has conducted groundbreaking work in this particular area [68, 69].

MSDs have been shown to be common among check-out workers, with work repetitiveness, especially since the introduction of laser scanners, a major contributing factor. The risk of MSDs is related to postural factors and the existence of generalized discomfort among workers. Laser scanners have indeed increased check-out workers' exposure to MSDs, to carpal tunnel syndrome in particular [70, 71].

Few studies involve both workers and management. One intervention implemented by an American supermarket that experienced increasing injury rates among check-out workers involved training programs for check-out workers, store managers, and management personnel, as well as major modifications to work stations and scanners. Injury rates were reduced and the company's management effort improved through increased awareness. Support from senior management and interdepartmental teamwork developed as an additional positive outcome [72].

Awkward postures are only one of several risk factors that may contribute to MSDs among supermarket check-out workers [73]. Repetitive handling of merchandise, for example, is another substantial source of physical strain to check-out workers. At check-out systems with laser scanners for sitting work posture, introducing job rotation where laser scanning is used is a useful work improvement technique [62]. Job rotation is a means of reducing workers' exposure to the repetitive movements used in laser scanning and the static posture of sitting at the check-out stand. Regular and frequent breaks, as well as task variation in a job, can be applied with benefits to airport check-in work as well (such as working with passengers with special needs, performing gate duty and safety functions), similar to other jobs characterized by static posture and repetitive work.

A rather unique and important approach has been to address the relationship between personal and occupational factors, including postures, movements, psycho-social factors at work, work station and store characteristics, and shoulder disorders [64]. While laser scanners have been shown to be an occupational risk factor, job control and wage dependence also have effects on efficiency. Psycho-social factors, such as monotonous work, workload, lack of development opportunities, job control, and social support reflect aspects of work organization that have negative health effects on supermarket check-out workers [74]. Shoulder disorders result from diverse work organization factors. Applying a systems perspective, recognizing the potential negative impacts on workers' health and well-being from personal, occupational, physical, psycho-social, and work organization factors, a much greater understanding of the relation between work and health can be achieved.

Where workers' control over the job is low (or non-existent), social support is even more important in protecting against ill health. Social support is essentially the glue that holds people together under adverse conditions.

Participatory ergonomics has demonstrated benefit in the case of computer users. Similar principles have been applied successfully to supermarket check-out work station re-design, with successful participatory re-design of supermarket check-out work stations involving workers in the workplace change process. As shown in the examples of participatory interventions with computer users, supermarket check-out workers are also a good source of information on workplace ergonomic problems and solutions. Successful participatory interventions emphasize the need for workplace health and safety, education, and interventions to focus on improving work conditions and work organization, and to be on-going [75, 76]. Evidence supports the benefits of participatory ergonomics as a strategy to improve health and safety [77] and also demonstrates links between psycho-social factors and MSDs and physical factors causing MSDs [78].

What is known to date about supermarket check-out workers reveals that a more expanded view is needed than what has constituted more traditional biomechanical research in this area. Characteristics of women's work, psycho-social factors, and organizational factors as potential determinants of MSDs need to be considered in addition to physical factors. The majority of studies have been rather limited in perspective and have focused mainly on physical factors for supermarket check-out workers. Stress resulting from management practices, psychological harassment, and emotional demands, for example, are known to contribute to muscular tension, making stress therefore an important contributor to MSDs, but which has been little considered. Applying a systems approach to an examination of carpal tunnel syndrome in supermarket check-out workers, for example, would yield a deeper understanding of the range of factors affecting workers. Looking at the job solely in relation to laser scanning and work station design is comparable to examining computer work solely from a keyboard and chair height perspective. Without considering job demands, level of control over one's own work organization, management practices, the existence of social support from co-workers and managers, and the impact of service-sector work, a full picture cannot be obtained of the factors that may contribute to, or help to prevent, a health problem that appears to be work-related in nature.

Studies that adopt a multi-variable and multidisciplinary approach to job examination (for both supermarket check-out workers and computer workers, e.g., [72, 75, 78] are more likely to report catalyzing a change process aimed at improvement for workers. Such studies report more positive outcomes for workers. The intervention-oriented studies on computer clerical workers and supermarket check-out workers argue convincingly for the examination of *psycho-social factors, physical factors, and organizational factors in combination* when looking at the causes of MSDs.

Finally, studies of supermarket check-out workers lack consideration of work environment factors. Elements as basic as temperature extremes or noise can have important consequences for muscular tension and psychological distress, particularly when experienced over the length of an entire work shift on a daily basis. Similarly, MSDs among check-in workers are aggravated by temperature extremes, at least in part caused by the constant opening and closing of airport doors, which usually face the workers. Would not supermarket check-out workers suffer from the same effects when they sit near or directly face automatic doors, as check-in workers so often do? Both groups may be subject to frequent blasts of cold or hot air, depending on the season, during an entire work shift. Such environmental factors may be direct or aggravating causes of MSDs in certain workers and should be taken seriously.

Airport Baggage Handlers

Below is a summary of what is known about airport baggage handlers, broken down by factors in common with airport check-in workers.

The majority of research conducted has focused only on MSDs in airport baggage handlers. Only one study identified [79] has taken a more expansive view, examining lack of control over job activities, work environment stressors, stress, and psycho-social issues among baggage handlers. There is, therefore, scope to examine broader issues, particularly those of an organizational nature, thought to impact baggage handlers.

Semi-mechanized airport check-in systems that require workers to lift and carry baggage have in common some of the same exposures, and therefore the potential for MSDs, as risks associated with airport baggage handlers. Recommendations made to protect baggage handlers, therefore, appear to have logical applicability to airport check-in workers as well. Yet findings from studies of baggage handlers have not been extended to include recommendations to protect check-in workers. This logical linkage has not been advanced through research on airport baggage handlers. Perhaps others have not considered that jobs that appear safe, clean, even glamourous, such as airport check-in work, could have the same exposures and risks as those inherent in a more obviously physically demanding job, such as airport baggage handling, a job performed primarily by men.

Workers vary in the techniques they use to adapt to weights that need to be lifted. More specifically, men and women use different techniques to handle weights [80]. Age and physical condition also affect the manner in which weights can be manipulated at work. Other than studies conducted on nurses, most occupational health research is conducted on male worker populations, while the problems associated with lifting weights are not sex-specific. The prevailing body of scientific literature (again, excluding nurses) gives a clear impression that problems associated with lifting weights are indeed specific to male workers. The musculoskeletal problems that *women* suffer from lifting weights on the job and why most occupational health research is conducted on male worker populations have been discussed by Messing [3, 9].

Occupational standards for lifting weights are generally determined by research conducted in laboratories. However, the real-life differences in the way individuals adapt to work demands may render laboratory simulations inapplicable to the types of situations that arise in real workplaces. In spite of the gap between laboratory and real-life, it is often the laboratory-based research that provides the scientific evidence on which standards are based.

MSDs have been demonstrated among airport baggage handlers, associated with both the cumulative and punctual effects of regular manual lifting of weights. With few exceptions, studies on airport baggage handlers have been limited to either back injury or risk of heart disease. Most of the studies fit a traditional occupational health perspective, in general, not examining additional factors that may contribute to back pain or injuries, such as systems of work organization and the impact of management practices.

Only one identified study noted that check-in workers systematically were not included in investigations of incidence and type of work-related accidents among baggage handlers and ground personnel, even though check-in workers are included in the category of "ground personnel" [81]. Baggage-handling *facilities* have been described from an engineering viewpoint, but with no substantive information on baggage-handling *workers* or their health and safety issues in relation to the baggage handling facilities [82].

Human characteristics and capacities relative to materials handling have been examined, particularly developed in one case study involving baggage-handling facilities at London's Heathrow Airport [83]. Management's concern about whether moving baggage from containers onto new conveyor belts was more likely to cause back injury than the existing design was addressed, with no clear associations between the design models and the occurrence of MSDs. An emphasis was put on the need for regulatory bodies to take greater account of the limitations of existing guidelines for lifting. The benefits of a commercially available weightlifting belt relative to reduced lumbar injury incident rate and injury severity among baggage handlers were examined by Reddell and colleagues [84]. Increased risk of injury was found when periods of no belt use followed periods of belt use, with belt use compliance emerging as an overriding factor. Weightlifting belts cannot be considered sufficiently protective if they are not used consistently, and therefore are not recommended for use by baggage handlers. This is one of the reasons that weightlifting belts have been mired in controversy.

Performance conditions and biomechanics have been used to classify how male workers at a German airport lifted and carried baggage [85]. Workers reported a range of MSDs, with low back pain causing prolonged periods of absenteeism. Sudden re-entry into former work activities was identified as a risk factor for back injury recurrence. The results led to training for the baggage handlers on lifting and carrying, using what was considered to be appropriate work technique. (It was one of the few studies on baggage handlers that resulted in an intervention. The intervention was entirely worker behavior--centered and involved no re-design of the work process.) Also applying a worker-centered focus, as compared with an organizational or work-process focus, Scheier examined training and education as a means of obtaining safety improvements for baggage handlers [86]. Worker-centered approaches place the burden of "change" on the worker, without addressing the systemic causes that invariably contributed to problems, such as design, work organization, and management practices.

Lifting, carrying, pulling, and pushing of loads in limited spaces have been found to promote back disorder in airport transport workers. Among workers operating in limited airplane compartment spaces, MSDs were common complaints, with back pain the most frequently reported (among 366 baggage handlers, 66% reported symptoms of back pain) [87, 88]. Other researchers have examined muscle strength in airport transport workers, without investigating MSDs [89], while still others measured postural forces during baggage handling at a large airport [90]. None of these studies made recommendations for improvements in working conditions. The United Kingdom's Health and Safety Executive (the national body for research and legislation in occupational health and safety) has provided guidance for injury prevention during baggage-handling operations at airports, based on the UK's 1992 Manual Handling Operations Regulations [91].

Several occupational health studies have examined factors beyond back injury alone. The range of chemical and physical exposures that can lead to morbidity among airport workers, compared with other work settings, has been studied based on the experiences of aviation ground personnel [92]. Prevention programs designed for low back pain and coronary artery disease were suggested, based on findings of significant rates of both among airport ground personnel. The influences of age, qualification, and shift type on baggage handlers' heart rates and described techniques and training for loading and unloading baggage have also been examined [93, 94]. The techniques used and training conducted were exclusively worker-centered.

Passenger and traffic services and baggage-handling facilities in Stockholm, Sweden were examined from a wider scope than that used by most researchers [80]. In addition to general and organizational factors, including management issues, physical environment, individual and group factors, work hours, and job design were examined. Work areas that emerged as priority concerns included computer workplace design with special functions, such as check-in counters; psychological and social environment, including management and supervisory issues; and the need to improve work conditions in aircraft baggage and cargo holds.

Conditions of work and work intensification for airline cabin crew were investigated in conjunction with the ITF [95], including examination of the physical work environment. Attention was paid to physical work environment factors, which are considered important in studying check-in workers and could be useful to examine among supermarket check-out workers, yet which were not addressed in any studies of the latter. An expanded view of the worker population was taken by recognizing the myths, or images, surrounding jobs associated with flying. Findings from the study dispel the myth of the flight attendant "glamour job" and reveal that cabin crew work in cramped, overcrowded, and unhealthy environments where they are under constant pressure to complete an increasing number of tasks within a tight time schedule. The "glamour job" image attached to flight attendants applies as well to check-in workers, and is suspected to explain, at least in part, the lack of attention to check-in work conditions.

Key findings of the Boyd and Bain study [95] undertaken in conjunction with the ITF included:

- Patterns of work are dominated by ultra-flexible shift work patterns and long duty hours. The majority of respondents said that they required more sleep and felt less healthy as a consequence.
- The majority of respondents were not entitled to a crew break during work. Almost two-thirds of respondents' work days lasted nine to twelve hours.
- The majority of respondents said that the intensity, speed, and volume of work had increased in only the last year, as had stress and pressure at work.
- The majority of respondents said that the number of abusive/disruptive passengers had increased in the last year alone. Respondents considered restraint-training courses inadequate.
- More than three-quarters of respondents felt that their health had deteriorated since starting their job. Factors such as cabin air quality, work patterns, poor hygiene, and close proximity to passengers were considered causes.
- Respondents' key health and safety concerns were cabin air quality, hygiene on aircraft, levels of stress and anxiety, and manual handling issues (pulling carts, lifting baggage, etc.).
- A high proportion of respondents had received no training in key areas such as manual handling, contact with body fluids (vomit, blood, saliva, etc.) or in dealing with violence
  - slightly more than a third of respondents took complaints to management, and only slightly more took complaints to the trade union.

It would make sense to extend to airport check-in workers the recommendations from studies of airport baggage handlers, which has not been done heretofore. In general, knowledge about exposures to men in one job may not appear to have direct use-value to a completely different job, particularly one performed by women. However, on closer examination, jobs that may appear completely different from the outside, in reality, may have numerous common characteristics and exposures from a variety of sources. Thus, knowledge about one job may indeed be useful to understand another that has not yet been investigated.

Studies pertaining to both computer clerical and supermarket check-out workers reveal a significant gap in the knowledge base about most occupations as they are performed and about women's jobs in particular. The gap could be narrowed by broadening the scope in both traditional occupational health and ergonomics studies to take into equal consideration jobs performed by women. Research on predominantly male occupations (in particular, airport baggage handlers) can serve as an information base for airport check-in work.

General Findings from Comparison Populations

Evidenced by research results about computer clerical workers, supermarket check-out workers, and airport baggage handlers, jobs that may appear completely different from the outside may, upon closer examination, have common exposures from a variety of sources. Thus, knowledge about one job may indeed be useful for another that has not been investigated. Similarities between these occupations leave little reason to doubt the existence of similar causes of MSDs reported by check-in workers.

Concluding from studies of these three comparison groups the mainstream research approach is more linear, or single factor, than multi-factorial, or holistic in nature. A multidisciplinary focus often is lacking where it could enhance the research. Indeed, based on the studies reviewed on these three groups, examining jobs from a systems approach is an exception rather than the norm. Among all three groups, a systems approach is a common denominator in the studies that led to workplace improvements and an organizational change process needed to sustain the improvements made. A study of check-in workers, examining multiple variables suspected to contribute to MSDs, should apply a similar approach, thereby attempting to take into account the totality of the work system.

The study conducted by Synnerman and Claridge supported these conclusions [30]. Physical as well as psycho-social factors were examined together with general work conditions. Priority concerns emerged that included computer workplace design, including check-in counters, psychological and social environment, including management and supervisory issues, and the need to improve work conditions. The Synnerman and Claridge study did not examine MSDs specifically, nor did it attempt to delineate causes and effects of any nature. Taking a multi-factorial view of check-in work necessarily means considering that more than one factor likely contributes to MSDs. Solutions should, therefore, follow in the same spirit and address numerous areas at once.

Observations, Outcomes

The study of check-in workers produced multiple findings. Here, I mention the most significant ones.

In the face of what can be difficult work conditions, check-in workers demonstrated positive attitudes toward their jobs and gave their best to ensure that passengers got off to their flights safely and contentedly. Check-in workers mostly reported that they enjoy the work they perform, enjoy working with the traveling public, enjoy working in airports, and often express a great sense of commitment to doing their best on the job [39]. In general, they also adopted a cooperative attitude toward management.

While they derive certain satisfactions from their work, the outcomes of the body mapping exercise revealed that many check-in workers live with pain in multiple body sites at once. Among the major groups of physiological and psycho-social stresses that have been identified by the U.S. National Institute for Occupational Safety and Health (NIOSH) as risk factors for work-related MSDs [96, 97], the study found six of them among the check-in workers, in varying degrees, at all three study sites:

1. Physiological stresses
   a. repetitiveness
   b. force and/or mechanical stress
   c. posture
2. Psycho-social stresses
   a. high demand
   b. low control
   c. violence

The workplace factors identified in the scientific literature as risks for MSDs (described above under the subheading *Perceptions aside, risks exist*), together with the level of mechanization of the baggage check-in system and a range of psycho-social factors, were confirmed as strong influences in the production of MSDs in airport check-in workers. Work station re-design, increased flexibility in work tasks, adjustable work station furniture, and workers' involvement in the decision-making process constitute solutions for reducing MSDs among check-in workers.

MSDs were prevalent, severe and caused disability among airport check-in workers. *More than 50%* of the workers experienced *severe neck and/or shoulder pain*, and *1 in 2* lived with *severe lower back pain*. Neck pain, shoulder disorders and lower back pain all resulted in *lost work time*. *"Presenteeism"* was significant: up to *¾ worked while suffering from MSD-related pain that impeded job performance*. *Over 70%* of workers paid a heavy price *sacrificing their sleep* to MSD-induced pain. Airport check-in work was revealed as an occupation likely to cause severe MSDs. Full mechanization of check-in systems is a sound investment for management and the savings from the costs of retrofitting check-in work stations or installing fully mechanized systems from the design stage outweigh the direct and indirect costs of MSDs to both workers and management. That change alone would have a notable mitigating influence on the level of MSDs in the workers.

Another major finding was the workers' concern about violence or the threat of violence on the job. *More than 80%* of the workers had been subjected to *verbal abuse* from passengers, *more than 20%* had received *threats* from passengers, and *1 in 20* had been *physically assaulted* on the job. To make matters worse, management support was noticeably absent when check-in workers had to deal with both aggressive passengers and the psychological distress and anxiety associated with violence on the job. And these were findings from two countries known for their peaceful and non-aggressive populations. (One must bear in mind that two of the study airports were international airports, where a mix of the entire world can constitute the traveling public.) Check-in workers are the first point of contact with passengers. They have to function no matter what emotional state passengers may display prior to travel. Since travel proves to be highly stressful for many people, a range of emotional states are displayed by passengers, including aggressive and violent reactions. Yet check-in workers seldom are part of an established system or hierarchy of measures designed to ensure passenger, crew, and aircraft safety against aggressive or potentially dangerous passengers. A re-focusing of roles would require management strategies to empower check-in workers and would involve policies, training, and organizational recognition of the preventive role in air transport safety that check-in workers could play.

The lack of training provided for workers was a clear and significant outcome. Airport check-in workers typically merely walk into their jobs. Most often they do not receive training for skills development, job enrichment, or worker protection, including training in ergonomics, safe lifting techniques, or self-protection against aggressive passengers. For the workers in this study, their only training focused on use of the computer software needed to perform the check-in job. In the area of psycho-social health, management practices were shown to be both direct and indirect causes of ill health in workers. The spillover effect into workers' personal lives from disempowering management practices cannot be dissociated from working life [98]. The link between direct and indirect causes of work-related ill health and the resulting impacts on private life has been revealed among check-in workers, where important numbers of workers reported that pain from MSDs interfered with sleep and caused them to stop activities outside of work. Management at airports around the world should realize that workers must be able to set parameters in the workplace. Increased autonomy and worker voice have been associated with increased benefits for both workers and management [34, pp. 105-129]. Flexibility in managerial attitudes is a rational and positive start toward participation in organizational decision-making.

Worker turnover is a key factor that essentially enables employers to externalize the costs of damage to workers' health. Check-in workers represent a de-professionalized sector of the civil aviation industry. Management policies increasingly target workers with seniority (generally those having higher salaries), encouraging them to leave their jobs so that management can hire younger workers who usually are paid lower wages, receive less attractive benefits, and sometimes have worse work conditions (for example, they may exclusively perform check-in work throughout an entire work shift rather than combine it with other tasks). Low wages, work pace, work load, and increase in short-term contracts or increasingly flexible work scheduling laws may be reasons for check-in workers to turnover more quickly now in some areas and to look elsewhere for better pay or better work conditions. Turnover though does not mean that conditions improve for newly hired workers. In areas where employment opportunities are limited, workers often accept poor work conditions and may be reluctant to demand or request improved conditions for fear of losing their job. In addition, management may use a slow economic period as an excuse for not improving conditions—a reminder that workers should be grateful to have a job no matter what the conditions.

The professional commitment of check-in workers is continuously undermined in the civil aviation industry. Sufficient recognition has not been accorded to the professional skills of check-in workers. The role of "safety professional on the ground" has not yet been widely linked with check-in work, despite the natural placement for this expanded function.

The study described in this book was the first ever undertaken on airport check-in workers and examined the impact of management practices on worker health [39]. But the health of check-in workers has not been visible on the management agenda. Worker health generally is not considered a priority by management of service-sector workers. Airport check-in workers are no exception. The lack of attention accorded to health is demonstrated by the lack of knowledge about such occupations and the extent of problems identified when such jobs are studied. Current prevalence rates and the persistent incidence of work-related accidents, injuries, illnesses, and fatalities demonstrate the low level of priority accorded to the protection of worker health in all sectors, in most workplaces, in most countries. Both Switzerland and Canada, where the study described in this book was carried out, have statutory regulations, and enforcement. But they are weak and inadequate. In many countries, including (and particularly) the United States, over the last 20 years statutory regulation has shifted to be dominated by "voluntary" initiatives often couched under enterprise-generated "safety culture" themes, which all too often is no more than a euphemism. Voluntary initiatives tend to be backed up by little or no enforcement and little threat of penalty for wrong doers [99]. Weak regulation plus lack of enforcement plus little threat of penalty is not a formula for strong protection. The result is that without statutory regulations, enforcement, financial incentives, and strong collective voice, workers' health remains conspicuously absent from management agendas.

Finally, this investigation raises questions about the ways in which job images are created, particularly when the image that is projected does not fit the objective conditions of work. While they may be seen or even see themselves as "glamour girls" in the airports, we now know that check-in workers are glamour girls living in pain and suffering.

ENDNOTES

1. ISCO-88 International Standard Classification of Occupations, International Labour Office, Geneva, 1990.
2. L. I. Boden, Workers' Compensation in the United States: High Costs, Low Benefits, *Annual Review of Public Health, 16,* pp. 189-218, 1995.
3. K. Messing, *One-Eyed Science: Occupational Health and Women Workers,* Temple University Press, Philadelphia, 1998.
4. L. Doyal, *What Makes Women Sick? Gender and the Political Economy of Health,* Macmillan Press, London, p. 153, 1995.
5. D. K. Wagener and D. W. Winn, Injuries in Working Populations: Black-White Differences, *American Journal of Public Health, 81,* pp. 1408-1414, 1991.
6. J. C. Robinson, Trends in Racial Equality and Inequality and Exposure to Work-Related Hazards, *American Association of Occupational Health Nurses Journal, 37,* pp. 56-63, 1989.
7. A. Pines, C. Lemesch, and O. Grafstein, Regression Analysis of Time Trends, *Occupational Accidents in Safety Science, 15,* pp. 77-95, 1992.
8. National Institute for Occupational Safety and Health, *Fatal Injuries to Workers in the United States, 1980-1989,* NIOSH, U.S. Department of Health and Human Services, 1993.
9. K. Messing (ed.), *Integrating Gender in Ergonomic Analysis: Strategies for Transforming Women's Work,* European Trade Union Technical Bureau for Health and Safety, Brussels, p. 53, 1999.
10. K. Messing, L. Dumais, and P. Romito, Prostitutes and Chimney Sweeps Both Have Problems: Towards Full Integration of Both Sexes in the Study of Occupational Health, *Social Science and Medicine, 36:*1, pp. 47-55, 1993.
11. E. Rosskam, Women Moving Mountains: Women Workers in Occupational Safety and Health, *Women and Environment International Journal, 48/49,* University of Toronto, pp. 18-20, 2000.
12. P. G. Dempsey, A. Burdorf, and B. S. Webster, The Influence of Personal Variables on Work-Related Low-Back Disorders and Implications for Future Research, *Journal of Occupational and Environmental Medicine, 39:*8, pp. 748-759, 1997.
13. K. Messing, L. Punnett, M. Bond, K. Alexanderson, J. Pyle, S. Zahm, D. Wegman, S. Stock, and S. de Grosbois, Be the Fairest of Them All: Challenges and Recommendations for the Treatment of Gender, *Occupational Health Research in American Journal of Industrial Medicine, 43,* pp. 618-629, 2003.
14. J. V. Johnson and E. M. Hall, Job Strain, Work Place Social Support and Cardiovascular Disease: A Cross-Sectional Study of a Random Sample of the Swedish Working Population, *American Journal of Public Health, 78:*10, pp. 1336-1342, 1988.
15. R. Karasek and T. Theorell, *Healthy Work: Stress, Productivity and the Reconstruction of Working Life,* Basic Books, New York, 1991.
16. R. Karasek, D. Baker, F. Marxer, A. Ahlbom, and T. Theorell, Job Decision Latitude, Job Demands and Cardiovascular Disease: A Prospective Study of Swedish Men, *American Journal of Public Health, 71:*7, pp. 694-705, 1981.
17. A discussion of emotional labor in check-in work is found in chapter 4.

18. E. Rosskam, Working at the Check-In: Consequences for Health, *New Solutions,* *15*:3. pp. 221-244, 2005.
19. Some useful discussions reflecting the growing interest in how social inequalities are embodied in gender, sex, race and ethnicity, and in socio-economic gradients are found in N. Krieger, Genders, Sexes, and Health: What are the Connections—And Why Does it Matter? *International Journal of Epidemiology, 32,* pp. 652-657, 2003; N. Krieger, J. T. Chen, P. D. Waterman, D. H. Rehkopf, and S. V. Subramanian, Race/Ethnicity, Gender, and Monitoring Socioeconomic Gradients in Health: A Comparison of Area-Based Socioeconomic Measures—The Public Health Disparities Geocoding Project, *American Journal of Public Health, 93,* pp. 1655-1671, 2003; L. D. Kuzbansky, N. Krieger, I. Kawachi, B. Rockhill, G. K. Steel, and L. F. Berkman, United States: Social Inequality and the Burden of Poor Health, in *Challenging Inequities in Health: From Ethics to Action,* T. Evans, M. Whitehead, F. Diderichsen, A. Bhuiya, and M. Wirth (eds.), Oxford University Press, Oxford, pp. 104-121, 2001; N. Krieger and G. Davey Smith, Bodies Count & Body Counts: Social Epidemiology & Embodying Inequality, *Epidemiologic Reviews, 26,* pp. 92-103, 2004.
20. D. Mclain, Responses to Health and Safety Risk in the Work Environment, *Academy of Management Journal, 38*:6, p. 1727, 1995.
21. J-F. Chanlat, Nouveaux Modes de Gestion, Stress Professionnel et Santé au Travail in L'homme à l'echine Plié, I. Brunstein (dir), Desclee de Brouwer, pp. 29-60, 2000.
22. E. M. Hall, Gender, Work Control and Stress: A Theoretical Discussion and an Empirical Test, *International Journal of Health Services, 19,* pp. 725-745, 1989.
23. R. Bourbonnais, B. Larocque, C. Brisson, M. Vézina, and D. Laliberté, *Environnement Psychosocial du Travail in Enquête Sociale et de Santé 1998* (2e édition), Institut de la Statistique, Québec, pp. 571-581, 1998.
24. B. Larocque, C. Brisson, and C. Blanchette, Cohérence Interne, Validité Factorielle et Validité Discriminante de la Traduction Française des Échelles de Demande Psychologique et de Latitude Décisionnelle de "Job Content Questionnaire" de Karasek, *Revue d'Épidémiologie et de Santé Publique, 96,* pp. 371-381, 1998.
25. K. De Koning and M. Martin (eds.), *Participatory Research in Health: Issues and Experiences,* Zed Books, London, 1996.
26. B. L. Hall, Participatory Research: An Approach for Change, *Convergence: An International Journal of Adult Education, 8*:2, p. 30, 1975 (in M. Elden, Sharing the Research Work: Participative Research and its Role Demands, in *Human Inquiry: A Sourcebook of New Paradigm Research,* P. Reason and J. Rowan (eds.), John Wiley and Sons, Chichester, p. 266, 1981).
27. E. Rosskam, A. Drewczynski, and R. Bertolini, *Service on the Ground: Occupational Health of Airport Check-In Workers,* International Transport Workers' Federation and International Labour Organization, London, 2005.
28. R. Karasek, *Control in the Workplace and Its Health-Related Aspects in Job Control and Worker Health,* S. L. Sauter, J. J. Hurrell, and C. Cooper (eds.), John Wiley and Sons, Chichester, pp. 129-159, 1989.
29. A. Hochschild, *The Managed Heart: Commercialization of Human Feeling,* University of California Press, San Francisco, 1983.

30. L. Punnett and U. Bergqvist, Musculoskeletal Disorders in Visual Display Unit Work: Gender and Work Demands, *Occupational Medicine—State of the Art Reviews, 14*:1, pp. 113-124, 1999.

31. L. Karlqvist, E. Wigaeus Tornqvist, M. Hagberg, M. Hagman, and A. Toomingas, Self-Reported Working Conditions of VDU Operators and Associations with Musculoskeletal Symptoms: A Cross-Sectional Study Focussing on Gender Differences, *International Journal of Industrial Ergonomics, 30*, pp. 277-294, 2002.

32. World Health Organization, Work with Visual Display Terminals: Psychosocial Aspects and Health, *Journal of Occupational Medicine, 31*, pp. 957-968, 1989.

33. World Health Organization, *Visual Display Terminals and Workers Health in WHO Offset Publication*, No. 99, Geneva, 1987.

34. See L. Karlqvist, A Process for the Development, Specification and Evaluation of VDU Work Tables, *Applied Ergonomics, 29*:6, pp. 423-432, 1998; R. F. Bettendorf, M. S. Hoffman, E. Adams, H. McLoone, and C. J. Purvis, *The OERC Framework for Understanding Upper-Extremity Musculoskeletal Disorders Research, Theory and Management*, paper presented at The National Ergonomics Conference and Exhibition, New York, May 1998; M. Smith, Psychosocial Aspects of Working with Video Display Terminals (VDTs) and Employee Physical and Mental Health, *Ergonomics, 40*:10, pp. 1002-1015, 1997; P. Carayon, C-L. Yang, and S-Y. Lim, Examining the Relationship Between Job Design and Worker Strain Over Time in a Sample of Office Workers, *Ergonomics, 38*:6, pp. 1199-1211, 1995; G. Westlander, The Full-Time VDT Operator as a Working Person: Musculoskeletal Work Discomfort and Life Situation, *International Journal of Human-Computer Interaction, 6*:4, pp. 339-364, 1994; P. Carayon, Job Design and Job Stress in Office Workers, *Ergonomics, 36*:5, pp. 463-477, 1993; P. Carayon-Sainfort, The Use of Computers in Offices: Impact on Task Characteristics and Worker Stress, *International Journal of Human-Computer Interaction, 4*:3, pp. 245-261, 1992; S. L. Sauter, L. M. Schleifer, and S. J. Knutson, Work Posture, Workstation Design, and Musculoskeletal Discomfort in a VDT Data Entry Task, *Human Factors, 33*:2, pp. 151-167, 1991; A. Statham and E. Bravo, The Introduction of New Technology: Health Implications for Workers, *Women and Health, 16*:2, pp. 105-129, 1990; P. Sainfort, Job Design Predictors of Stress in Automated Offices, *Behaviour and Information Technology, 9*:1, pp. 3-16, 1990; S. Sauter, M. Dainoff, and M. Smith (eds.), *Promoting Health and Productivity in the Computerized Office*, Taylor and Francis, London, 1990; S. L. Sauter, M. S. Gottlieb, K. C. Jones, V. N. Dodson, and K. M. Rohrer, Job and Health Implications of VDT Use: Initial Results of the Wisconsin-NIOSH Study, *Human Aspects of Computing, 26*:4, pp. 284-294, 1983; M. J. Smith, B. G. F. Cohen, and L. W. Stammerjohn Jr., An Investigation of Health Complaints and Job Stress in Video Display Operations, *Human Factors, 23*:4, pp. 387-400, 1981.

35. C-N. Ong, S-E. Chia, J. Jeyaratnam, and K-C. Tan, Musculoskeletal Disorders Among Operators of Visual Display Terminals, *Scandinavian Journal of Work, Environment and Health, 21*, pp. 60-64, 1995.

36. J. Faucett and D. Rempel, VDT-Related Musculoskeletal Symptoms: Interactions Between Work Posture and Psychosocial Work Factors, *American Journal of Industrial Medicine, 26,* pp. 597-612, 1994.

37. A. C. Matias, G. Salvendy, and T. Kuczek, Predictive Models of Carpal Tunnel Syndrome Causation Among VDT Operators, *Ergonomics, 41:*2, pp. 213-226, 1998.

38. H. Shahnavaz, J. Abeysekera, and A. Johansson, Solving Multi-Factorial Work Environment Problems Through Participation: Case Study: VDT Operators, in *The Ergonomics of Manual Work,* W. Marras, W. Karwowski, J. Smith, and L. Pacholski (eds.), Taylor and Francis, London, pp. 499-502, 1993.

39. E. Rosskam, *Working at the Check-In: Consequences for Worker Health and Management Practices,* University of Lausanne, Lausanne, 2003.

40. M. J. Dainoff, Occupational Stress Factors in Visual Display Terminal (VDT) Operation: A Review of Empirical Research, *Behaviour and Information Technology—BIT, 1:*2, pp. 141-176, 1982.

41. M. Marcus and F. Gerr, Upper Extremity Musculoskeletal Symptoms Among Female Office Workers: Associations with Video Display Terminal Use and Occupational Psychosocial Stressors, *American Journal of Industrial Medicine, 29,* pp. 161-170, 1996.

42. K. Messing (ed.), *Integrating Gender in Ergonomic Analysis: Strategies for Transforming Women's Work,* European Trade Union Technical Bureau for Health and Safety, Brussels, 1999.

43. K. Messing, *One-Eyed Science: Occupational Health and Women Workers,* Temple University Press, Philadelphia, 1998.

44. K. Messing, C. Chatigny, and J. Courville, Light and Heavy Work in the Housekeeping Service of a Hospital, *Applied Ergonomics, 29:*6, pp. 451-459, 1998.

45. C. N. Ong, S. E. Chia, J. Jeyaratnam, and K. C. Tan, Technological Change and Work-Related Musculoskeletal Disorders: A Study of VDU Operators, in *Towards Human Work: Solutions to Problems in Occupational Health and Safety,* M. Kumashiro and E. Megaw (eds.), Taylor and Francis, London, pp. 333-339, 1991.

46. A. J. Haufler, M. Feuerstein, and G. D. Huang, Job Stress, Upper Extremity Pain and Functional Limitations in Symptomatic Computer Users, *American Journal of Industrial Medicine, 38,* pp. 507-515, 2000.

47. I. T. S. Yu, and T. W. Wong, Musculoskeletal Problems Among VDU Workers in a Hong Kong Bank, *Occupational Medicine, 46:*4, pp. 275-280, 1996.

48. L. Krapac and D. Sakic, Locomotor Strain Syndrome in Users of Video Display Terminals, *PubMed, 45:*4, pp. 341-347, 1994.

49. M. Ala, M. Ala, and G. Bagot, Ergonomics in the Laboratory Environment, *Nursing Management, 25:*7, pp. 50-52, 1994.

50. P. K. Nielsen, J. Andersen, and K. Jorgensen, The Muscular Load on the Lower Back and Shoulder Due to Lifting at Different Heights and Frequencies, *Applied Ergonomics, 29:*6, pp. 445-450, 1998.

51. U. Bergqvist, E. Wolgast, B. Nilsson, and M. Voss, The Influence of VDT Work on Musculoskeletal Disorders, *Ergonomics, 38,* pp. 754-762, 1995.

52. C. Wilholm and B. Arnetz, Musculoskeletal Symptoms and Headaches in VDU Users—A Psychophysiological Study, *Work and Stress, 11:*3, pp. 239-250, 1997.

53. T. R. Hales, S. L. Sauter, M. R. Peterson, L. J. Fine, V. Putz-Anderson, L. R. Schleifer, T. T. Ochs, and B. P. Bernard, Musculoskeletal Disorders Among Visual Display Terminal Users in a Telecommunication Company, *Ergonomics, 37*:10, pp. 1603-1621, 1994.
54. M. J. Smith and P. C. Sainfort, A Balance Theory of Job Design for Stress Reduction, *International Journal of Industrial Ergonomics, 4*, pp. 67-79, 1989.
55. A. G. Ryan, The Prevalence of Musculo-Skeletal Symptoms in Supermarket Workers, *Ergonomics, 32*:4, pp. 359-371, 1989.
56. M. A. Ayoub, Ergonomic Deficiencies: I. Pain at Work, *Journal of Occupational Medicine, 32*:1, pp. 52-57, 1990.
57. T. J. Armstrong, R. G. Radwin, D. J. Hansen, and K. W. Kennedy, Repetitive Trauma Disorders: Job Evaluation and Design, *Human Factors, 28*:3, pp. 325-336, 1986.
58. C. Carrasco, N. Coleman, S. Healey, and M. Lusted, Packing Products for Customers: An Ergonomics Evaluation of Three Supermarket Checkouts, *Applied Ergonomics, 26*:2, pp. 101-108, 1995.
59. B. Das and A. Sengupta, Industrial Workstation Design: A Systematic Ergonomics Approach, *Applied Ergonomics, 27*:3, pp. 157-163, 1996.
60. K. Grant, D. J. Habes, and S. L. Baron, An Ergonomics Evaluation of Cashier Work Activities at Checker-Unload Workstations, *Applied Ergonomics, 25*:5, pp. 310-318, 1994.
61. A. Johansson, G. Johansson, P. Lundqvist, I. Akesson, P. Odenrick, and R. Akselsson, Evaluation of a Workplace Redesign of a Grocery Checkout System, *Applied Ergonomics, 29*:4, pp. 261-266, 1998.
62. U. Hinnen, T. Laubli, U. Guggenbuhl, and H. Krueger, Design of Check-Out Systems Including Laser Scanners for Sitting Work Posture, *Scandinavian Journal of Work, Environment and Health, 18*:3, pp. 186-194, 1992.
63. M. Vézina, L. Geoffrion, C. Chatigny, and K. Messing, A Manual Materials Handling Job: Symptoms and Working Conditions Among Supermarket Cashiers, *Chronic Diseases in Canada, 15*:1, pp. 17-22, 1994.
64. I. Niedhammer, M. F. Landre, A. LeClerc, F. Bourgeois, P. Franchi, J. F. Chastang, G. Marignac, P. Mereau, D. Quinton, C. R. Du Noyer, A. Schmaus, and C. Vallayer, Shoulder Disorders Related to Work Organization and Other Occupational Factors Among Supermarket Cashiers, *International Journal of Occupational and Environmental Health, 4*:3, pp. 168-178, 1998.
65. D. Orgel, M. Milliron, and L. Frederick, Musculoskeletal Discomfort in Grocery Express Checkstand Workers: An Ergonomic Intervention Study, *Journal of Occupational Medicine, 34*:8, pp. 815-818, 1992.
66. S. Toon, Service Industries Say Safety Plays Key Role in Putting Customer First, *Occupational Health and Safety, 62*:11, pp. 61-69, 1993.
67. A. Soares, Tears at Work: Gender, Interaction and Emotional Labour, *Just Labour: A Canadian Journal of Work and Society, 2*, pp. 36-44, 2003.
68. A. Soares, Violence dans le Jardin d'Eden: Le Travail des Caissières au Brésil et au Québec, *Cahiers du Genre, 28*, pp. 97-115, 2000.
69. A. Soares, Les Qualifications Invisibles dans le Secteur des Services: Le Cas des Caissières de Supermarchés, *Lien Social et Politiques—RIAC, 40*, pp. 105-116, 1998.

70. P. Harber, D. Bloswick, J. Luo, J. Beck, D. Greer, and L. F. Pena, Work-Related Symptoms and Checkstand Configuration: An Experimental Study, *American Industrial Hygiene Association Journal, 54*:7, pp. 371-376, 1993.

71. W. Margolis and J. F. Kraus, The Prevalence of Carpal Tunnel Syndrome Symptoms in Female Supermarket Checkers, *Journal of Occupational Medicine, 29*:12, pp. 953-956, 1987.

72. E. Allen, Keeping Grocery Checkout Lines Moving, *Risk Management, 45*:1, pp. 38-39, 1998.

73. K. Grant and D. Habes, An Analysis of Scanning Postures Among Grocery Cashiers and Its Relationship to Checkstand Design, *Ergonomics, 38*:10, pp. 2078-2090, 1995.

74. J. R. Wilson and S. M. Grey, Reach Requirements and Job Attitudes at Laser-Scanner Checkout Systems, *Ergonomics, 27*:12, pp. 1247-1266, 1984.

75. I. Randle and A. Nicholson, Cashier Workstation Audit and Re-Design for Halifax PLC, *The Safety and Health Practitioner, 17*:7, pp. 50-52, 1999.

76. D. Orgel, M. Milliron, and L. Frederick, Musculoskeletal Discomfort in Grocery Express Checkstand Workers: An Ergonomic Intervention Study, *Journal of Occupational Medicine, 34*:8, pp. 815-818, 1992.

77. P. C. Bohr, B. A. Evanoff, and L. D. Wolf, Implementing Participatory Ergonomics Teams Among Health Care Workers, *American Journal of Industrial Medicine, 32*, pp. 190-196, 1997.

78. K. Fredriksson, L. Alfredsson, C. B. Thorbjornsson, L. Punnett, A. Toomingas, M. Torgen, and A. Kilbom, Risk Factors for Neck and Shoulder Disorders: A Nested Case-Control Study Covering a 24-Year Period, *American Journal of Industrial Medicine, 38*, pp. 516-528, 2000.

79. J. E. Synnerman and N. Claridge, The Work Environment Conditions of Civil Aviation Ground Staff, in *Ergo Management AB*, Stockholm, p. 238, 1984.

80. J. Stevenson, D. R. Greenhorn, J. T. Bryant, J. M. Deakin, and J. T. Smith, Gender Differences in Performance of a Selection Test Using the Incremental Lifting Machine, *Applied Ergonomics, 27*:1, pp. 45-52, 1996.

81. J. Ribak, B. Cline, and P. Froom, Common Accidents Among Airport Ground Personnel, *Aviation, Space, and Environmental Medicine, 66*:12, pp. 1188-1190, 1995.

82. Air Transport Association of America, *Report of the Air Transport Association Facility Planning Guidelines: Baggage Handling, Passenger Loading Bridges, Preconditioned Air Systems, Ground Power Systems*, BNP Associates, Inc., Connecticut, September 1998.

83. P. W. Buckle, D. A. Stubbs, I. P. M. Randle, and A. S. Nicholson, Limitations in the Application of Materials Handling Guidelines, *Ergonomics, 35*:9, pp. 955-964, 1992.

84. C. R. Reddell, J. J. Congleton, R. D. Huchingson, and J. F. Montgomery, An Evaluation of a Weightlifting Belt and Back Injury Prevention Training Class for Airline Baggage Handlers, *Applied Ergonomics, 23*:5, pp. 319-329, 1992.

85. A. Ruckert, W. Rohmert, and G. Pressel, Ergonomic Research Study on Aircraft Luggage Handling, *Ergonomics, 35*:9, pp. 997-1012, 1992.

86. R. Scheier, Airlines Confront a Ground-Level Safety Problem, *Safety and Health, 144*:1, pp. 64-68, 1991.

87. K. Undeutsch, R. Kupper, I. Lowenthal, K. H. Gartner, T. Luopajarvi, K. Rauterberg, M. J. Karvonen, and J. Rutenfranz, Occupational Health Studies on Airport Transport Workers—III. Musculoskeletal Complaints and Orthopedic Disorders of Airport Transport Workers, *International Archives of Occupational and Environmental Health, 50*:1, pp. 59-75, 1982.

88. K. Undeutsch, K. H. Gartner, T. Luopajarvi, R. Kupper, M. J. Karvonen, I. Lowenthal, and J. Rutenfranz, Back Complaints and Findings in Transport Workers Performing Physically Heavy Work, *Scandinavian Journal of Work Environment Health, 8*:1, pp. 92-96, 1982.

89. M. J. Karvonen, J. Mainzer, W. Rohmert, I. Lowenthal, K. Undeutsch, R. Kupper, K. H. Gartner, and J. Rutenfranz, Occupational Health Studies on Airport Transport Workers—II. Muscle Strength of Airport Transport Workers, *International Archives of Occupational and Environmental Health, 47*:3, pp. 233-244, 1980.

90. W. Diebschlag, F. Heidinger, B. Kurz, I. Lowenthal, W. Rohmert, and A. Ruckert, Measurement of Actual Forces During Luggage and Pallet Handling at a Large Airport, *Arbeitsmedizin–Sozialmedizin–Praventivmedizin, 24*:5, pp. 98-105, 1989.

91. Health and Safety Executive, Manual Handling Operations: Baggage Handling at Airports in HSE Books, Sudbury, United Kingdom, September 1996.

92. P. Froom, B. Cline, and J. Ribak, Disease Evaluated on Return-to-Work Examinations: Aviation Ground Personnel Compared to Other Workers, *Clinical Medicine, 67*:4, pp. 361-363, 1996.

93. W. Rohmert, I. Lowenthal, and A. Ruckert, Workload and Physiological Strain Caused by Baggage Handling at a Large Airport, *Arbeitsmedizin–Sozialmedizin–Praventivmedizin, 24*:3, pp. 47-52, 1989.

94. A. Ruckert and W. Rohmert, Evaluation of Manual Load Handling in Confined Spaces, *Arbeitsmedizin–Sozialmedizin–Praventivmedizin, 26*:2, pp. 58-63, 1991.

95. C. Boyd and P. Bain, *A Summary of the Findings from the Cabin Crew Health and Working Environment Survey,* University of Strathclyde, United Kingdom in collaboration with Transport and General Workers Union and BASSA, 1999.

96. National Institute for Occupational Safety and Health, *Cumulative Trauma Disorders in the Workplace,* NIOSH, U.S. Department of Health and Human Services, NIOSH Publication No. 95-119, 1995.

97. S. Snook, The Practical Application of Ergonomics Principles, in *Minesafe International,* 1993.

98. J-F. Chanlat, Stress, Psychopathologie du Travail et Gestion, in *L'Individu dans l'Organization: Les Dimensions Oubliées,* J-F. Chanlat (ed.), Presses de l'université Laval/Eska, Sainte Foy, Paris, pp. 709-721, 1990.

99. This is not to say that there are not many excellent enterprises taking seriously the protection of workers' health. Those that are leading examples based on their own "voluntary initiative" should be lauded and widely publicized as desirable employers, examples for others to follow.

CHAPTER 3

Strong Backs, Strong Shoulders:
So Why Musculoskeletal Disorders?

Musculoskeletal disorders have become the scourge of today's workplace. Simply having a strong back and strong shoulders does not ward off musculo-skeletal disorders induced by an ill-conceived job, especially if the job is combined with management practices that damage workers' health. Only the tip of the iceberg of MSDs is known and only in a handful of countries.

Research on MSDs has tended to focus on physical factors as inducers of MSDs. The existing body of literature, however, underscores the need to focus increasingly on their *multi-factorial etiology* and to give special attention to the contribution of management practices and psycho-social factors to their development and severity. The current prevailing (and limited) research perspective that dominates in particular the body of North American and occupational health literature is paradoxically an important contributor to the prevalence of MSDs, their increase, and their standing as a recognized major problem for workers, employers, insurance companies and society in general. When research focuses primarily or exclusively on physical factors, biomechanics, or design factors, both standard-setting and workplace-based action will tend to do the same, leaving other critical factors (particularly cognitive and psycho-social factors) un-addressed in the very many jobs where physical factors are only one piece of the causality picture. Francophone ergonomics tends to differ from Anglophone ergonomics, the former promoting a systems approach, while North American ergonomics research has tended to concentrate [often exclusively] on bio-mechanics, physical factors, or job tasks through work analysis, often omitting cognitive and psycho-social factors [1].

In his early 18th century book, *De Morbus Artificum*, Bernardino Ramazzini described illnesses caused by "violent and irregular motions and unnatural postures of the body" [2]. Ramazzini anticipated the widespread occurrence of MSDs in modern white-collar workplaces when he wrote,

> The diseases . . . arise from three causes: first constant sitting, the perpetual motion of the mind . . . constant writing also considerably fatigues the hand and the whole arm on account of the continual and almost tense tension of the muscles and tendons [B. Ramazzini, De Morbus Artificum, 1706, in Europe Under Strain, R. O' Neill, European Trade Union Technical Bureau for Health and Safety, Brussels, 1999, p. 31].

Nearly 300 years later, the same risk factors cause the same diseases in workers, while one important factor has changed: new management techniques and automation have been introduced to improve productivity and profitability, intensifying work and creating significant degrees of strain, psychological distress and pressure on workers. If these risk factors and diseases are the same as those noted in the 18th century, the knowledge gained has not been applied universally to the improvement of work organization and work station design, demonstrated by the prevalence rates and costs of MSDs [2]. The burden of these disorders is great, but the problems can be reduced with well-designed intervention programs and changes in management practices.

The European Agency for Safety and Health at Work has estimated that at least 40% of European workers are affected by MSDs [3, 4]. The 1995 Eurostat study, launched by the European Commission to achieve comparability of data on recognized occupational diseases in member states, indicated that MSDs were among the ten most frequent diseases in the European Union, with upper limb disorders in sixth and seventh position [5]. The January 1996 survey of work conditions by the European Foundation for the Improvement of Living and Working Conditions revealed that almost one in every five workers in the EU work force took time off work during the previous year due to health problems caused by their job.

A third of workers undertook jobs involving manipulating heavy loads. Thirty percent suffered from backache and 45% had painful or tiring work positions. More than half of the workers had to perform repetitive hand and arm movements or do their work at a high rhythm. One-third had no influence on their work rhythms and methods and two-thirds had their pace of work determined by customers, patients, passengers, or other service users. The study concluded that the inability of workers to take breaks when they wished, control the pace of work, and decide about the organization of their work are key risk factors for MSDs.

Indeed these figures for European workers are high by any account. But consider them in comparison with this study's findings on the suffering that airport check-in workers live with: *more than 70%* of the respondents indicated that they live with neck pain severe enough to affect their work performance. Among check-in workers, MSDs are not only common but can lead to temporary or permanent disability. Other types of musculoskeletal pain experienced and degree of severity are described later in this chapter.

The Nederlands Instituut voor Arbeidsomstandigheden (NIA) generated national level results similar to those of the European Foundation. Findings showed that about 30% of workers in the Netherlands—two million workers—are at risk of repetitive strain injuries (RSI), a musculoskeletal disorder caused by repetitive movements. One in every five Dutch workers reported problems with the neck, shoulders, arms, wrists, or hands to their company doctor, and 4% of all incapacity for work was caused by similar complaints [6].

MSDs of the low back and upper extremities have been described as a significant health problem in the United States as well, resulting in approximately one million people losing time from work each year [7]. Work-related MSDs cause more lost work time than any other occupational health problem and associated compensation and insurance costs are higher than for most workplace health problems. For cases involving days away from work in 1994, approximately 705,800 cases (32%) were the result of overexertion or repetitive motion [8]. Specifically, there were:

- 367,424 injuries due to overexertion in lifting (65% affected the back); 93,325 injuries due to overexertion in pushing or pulling objects (52% affected the back); 68,992 injuries due to overexertion in holding, carrying, or turning objects (58% affected the back). Totaled across these three categories, 47,861 disorders affected the shoulder;
- 92,576 injuries or illnesses due to repetitive motion, including typing or key entry, repetitive use of tools, and repetitive placing, grasping, or moving of objects other than tools. Of these ailments, 55% affected the wrist, 7% affected the shoulder, and 6% affected the back;
- 83,483 injuries or illnesses in other and unspecified overexertion events.

Not stagnating in the working population, the number of repeated trauma cases rose steadily from 23,800 in 1972, to 332,000 in 1994—a 14-fold increase [8] without taking into consideration the significant proportion of MSDs that might be work-related but were never reported as such. Morse, Dillon, Warren, Levenstein, and Warren estimated that approximately one in ten working-age adults in the United States suffers from a work-related MSD of the neck, arm, wrist, or hand, 20% of which are expected to have been physician-diagnosed, but only 10% of which had been compensated as work-related [9]. This information suggests a very high rate of underreporting. Often hidden, the costs to workers' personal lives from work-related MSDs can be significant, including losses of homes and cars, divorces, and job dislocations.

A 1994 French Ministry of Labor survey concluded that worldwide, 3.4 million workers were exposed to MSDs, estimated to represent 28% of the global labor force. Thirteen percent of the world's workers were estimated as exposed to repetitive work at high speed, 8% exposed to work in static postures, and 7% exposed to both. MSDs are the leading workplace hazard in European workplaces.

The burden is high, and the available information may represent only the tip of the iceberg for industrialized countries. If data were available from transitional economies and developing countries, unquestionably the extent of suffering would be demonstrated as far, far greater, given the extent of heavy physical work under extreme conditions that prevails in many countries, compounded by stress induced by poor management practices and insecurity-induced psychological distress.

CAUSES OF MUSCULOSKELETAL DISORDERS

A substantial body of credible epidemiological research provides strong evidence of an association between MSDs and multi-factorial work-related physical factors when there are high levels of exposure, especially in combination with exposure to more than one physical factor such as repetitive lifting of heavy objects in extreme or awkward postures. A large population-based survey in Québec found that working people who work standing and/or in constrained postures are more likely to be exposed to other physical work demands (e.g., handling heavy loads, repetitive work, forceful exertion and low job decision latitude) [10]. Among workers usually constrained to sitting at work, static, constrained postures were associated with lower income and educational levels, and with low-decision latitude, consistent with characteristics of jobs performed predominantly by women. These findings correspond with the picture discovered in this study of check-in workers, where multiple physical work demands co-exist with constrained postures identified in two of the three study site airports, combined with low job decision latitude and low wage work at all three sites.

Scientific support for the concept of a multi-factorial etiology of work-related low back pain has been demonstrated through examination of low back pain among industrial and municipal waste workers [11, 12]. Yet despite the fact that low back disorders continue to be one of the largest single sources of workers' compensation costs, the contribution of personal, workplace, organizational, and environmental variables to low back disorders is not well understood. While many studies exist involving all of these factors, they are not well understood because typically the various factors are looked at one by one, and because one individual factor occurring in one situation often corresponds to a different reality elsewhere, making comparisons difficult.

Occupational risk factors of work load and exposure time in jobs involving manual materials handling are particularly relevant to airport check-in workers. The commonality of exposure factors emphasizes the need to analyze check-in work from a holistic perspective, as for any other job involving manual materials handling.

NIOSH in the United States described physical stressors and workplace conditions that pose a risk of injury or illness to a worker's musculoskeletal system as the main characteristics of jobs and pressures leading to musculoskeletal strains in workers. Framed within these two major job characteristics leading to MSDs, NIOSH defined five key environmental hazards as:

- Repetitive motions
- Forceful motions
- Vibration
- Temperature extremes
- Awkward postures that arise from improperly designed work stations, tools, and equipment; and improper work methods [13].

Intensity and length of exposure to forceful, repetitive work have been identified as additional risk factors [14].

Work-related physical stressors are found to contribute to a considerable portion of work-related low back disorders [13, 15].

> The effects of ergonomic hazards may be amplified by extreme environmental conditions. In addition, ergonomic hazards may arise from potentially deleterious job designs and organizational factors such as: excessive work rates; external (versus self) pacing of work; excessive work duration; shift work; imbalanced work-to-rest ratios; demanding incentive-pay or work standards; restriction of operator body movement and confinement of the worker to a work station without adequate relief periods; electronic monitoring; and lack of task variety [2, p. 18].

Accounting for many MSDs and repetitive strain injuries is the increase in work intensity, created by modern technological systems and management practices designed to maximize their use, causing an erosion of terms and conditions of employment, and violating workers' fundamental rights [16]. MSDs and repetitive strain injuries have been shown to increase when work time rises [17]. A study of stress and recurrent low back pain among Finnish planners, welders, and plumbers (all male) found that older planners reported the most physical exhaustion and musculoskeletal problems. Problems in work organization were strongly associated with stress, weak employment security, and low back pain, and low back pain increased with age among manual workers whose work was characterized by noise, heavy lifting, awkward postures and high demands [18]. All of these characteristics are found in work performed by airport check-in workers.

In Europe, work intensification has increased to an important extent. Six years ago, 54% of workers worked at very high speed and 56% worked under the pressure of tight deadlines. The 2000 European survey of the European

Foundation revealed that 56% of workers worked at very high speed, 60% of workers worked under the pressure of tight deadlines, and 15% of tasks in European work environments had a work cycle of less than five seconds, which is considered highly repetitive [19].

In summary, the European Foundation's 1996 Survey revealed that work in the European Union has the following characteristics associated with MSDs:

- Excessive loads and poor design of workplaces;
- Fast pace of work, repetitive work, work patterns not designed for the workers performing the job, payment systems related to productivity, and boring, monotonous work;
- An increasing intensity of work—faster speed, tighter deadlines, and just-in-time policies;
- Lack of worker autonomy, including the inability to take breaks when needed, to control the pace of work, and to influence the organization of work;
- The increasing precariousness of work leading to difficulty resisting pressure to overuse.

The study also shows more women workers are employed in jobs carrying a risk of MSDs than males; and that risk of MSDs exists in all industrial sectors [20].

While traditional risk factors continue to affect a high proportion of workers, and are not decreasing, new risk factors linked to organizational developments are increasing. In particular, work is getting more intense and work patterns are becoming more irregular, with work-induced stress inextricably linked to MSDs [21], and job satisfaction significantly negatively affected by the conditions of work. Confirming the *reciprocal* causal relation between stress and MSDs, Leino [22] found that stress symptoms were associated with rheumatic symptoms and clinical findings of chronic MSDs among male workers and skilled and semiskilled blue collar female workers. An important study in the literature, the findings revealed that MSDs produce stress in skilled and semiskilled workers *and* that stress produces MSDs in the same workers.

> Degeneration of the locomotor system and the associated symptoms interfere with movement and cause difficulties in pursuing everyday tasks, etc. They are, therefore, sources of stress. . . . [and] stress increases the occurrence of musculoskeletal disorders. Some of the effect may be mediated by influences on help seeking behavior, but the results also suggest that stress may produce changes in the physical state of the musculoskeletal system [22, p. 299].

Static loads on the muscles associated with awkward postures, prolonged sitting, and prolonged standing, all of which are found in airport check-in work, are recognized causes of fatigue and contribute to MSDs [23]. Good workplace design is essential to avoid excessive fatigue and to help prevent MSDs. And attention to good workplace design—*from a holistic perspective*—is essential for job satisfaction, which has been more strongly associated with psycho-social factors than with physical complaints. The statistical correlation between job satisfaction and health bears very important implications for worker health and well-being and workplace interventions [24].

As part of any discussion of the impacts of static load and MSDs in general, the particular problems of women workers deserve special attention since around the world, women workers are the most exposed to static postures, with a higher prevalence among low and unskilled workers [25].

WOMEN WORKERS, WOMEN'S WORK, AND MSDs

In most countries of the world, women are concentrated in particular sectors of the economy—service jobs, selected areas of manufacturing, and in agriculture. Within each of these areas of work, women are concentrated in the jobs with the lowest pay, the least status, often involving monotonous and repetitive work [26, 27].

Only in the 1970s did any systematic investigation begin of the health effects of women's work [28, 29]. Still today, most work-related research focuses on major industries in industrialized countries. The majority of research focuses on traditional male jobs well recognized for workers' compensation [30]. Little examination has been undertaken recognizing the multi-causal relationships and adverse effects on women's health from precarious employment and the particular vulnerability of women in the labor market. Thus, what is known about women's work and its effects on women workers' health is relatively limited.

This is true despite the fact that many jobs performed exclusively or mostly by women have an important physical component, which can produce pain and even disability. When women workers have work-related health problems, mostly they are attributed to personal weakness and psychological difficulties, but the same reasoning is seldom applied to men's occupational health issues. The misguided assumption that women's work is "safe" means also that few prevention programs exist for women workers.

Research in the Nordic countries by Aittomaki, Lahelma, Leino-Arjas, and Martikainen [31] showed that strong gender differences in physical workload among male and female workers have been associated with increasing age, where physical workload *decreases for men as they get older, but not for women, particularly women with low socio-economic status*. Findings showed that the most physically demanding occupations were less common in men over 55 years old, while corresponding differences were modest for women. Such demands on older women workers are linked to difficulty in carrying out and completing daily tasks and can lead to increases in MSDs among women workers. Similar findings have been previously shown by Torgén and Kilbom [32].

Disentangling the influence of gender from other factors in workplace cross-sectional studies is a difficult undertaking, one that warrants pursuit in future research but which, to date, has been tackled very little. Yet the obvious differences in physical capacities between men and women are a strong argument for careful consideration of the gender influences in epidemiologic studies [15, p. 750]. Messing and Elabidi brought to light the power of perceptions and assumptions created by gender-based task assignment and the need for observation and discussion to accompany the desegregation of jobs where it occurs [33].

The need to consider the gender influences in MSDs is so obvious that only the socio-political forces at work explain the lack of research in this area [34]. The lack of research on women's occupational health is also reflected in the methods used to determine appropriate standards for physical work conditions, such as temperature as well as maximum weights to be lifted. For workplace standards and the research on which such standards are based to be appropriate for women as well as men, they need to take into account the physiology, anatomy, and anthropometry of women, the physical stressors women experience at work, as well as those they return to at home. This approach is indispensable because the average woman is a different size and shape from the average man. In typically male-dominated jobs, the workplace is designed for male anthropometry and male norms, including most work tables and work stations where women work [34].

Standards based on men's physiology and anthropometry will not protect most women. And further differences must be taken into account between male and female size and shape in different countries and among different races. Taking them into consideration is particularly important in this era of globalization, where not only tools but entire work stations are exported from one part of the world to another, even though they may be entirely ill-suited to workers in the receiving country. Flexibility and adjustability in design and materials are, therefore, of paramount importance. This approach applies equally to airport check-in work, where a good work station fit is needed for both male and female workers, allowing for anthropometrical differences found among workers in different countries.

In North America and Europe, women workers are disproportionately employed in work that is monotonous, repetitive, has a high level of job strain, demands a low level of skill, and that is considered low-wage. The combination of these work characteristics has a high risk for MSDs. A variety of physical injuries is common in what is typically women's work—nursing aides, cashiers, poultry workers, sewing machine operators [35] (U.S. Bureau of Labor Statistics [8]; European Commission [6; European Foundation for the Improvement of Living and Working Conditions [20]), aircraft cabin cleaners [36], and hotel room cleaners [37, 38]. Work-related MSDs have been well studied because they constitute the majority of cases of occupational disease today, and work-related MSDs are also the most common of women's work-related health problems [39].

Recent studies of aircraft cabin cleaners in Hong Kong [36] and hotel room cleaners in Las Vegas [37]—both jobs performed predominantly by women—revealed high levels of MSDs, pain, and injury among workers. Both jobs were shown to entail extremely ill conceived tasks; workers performed daily functions under the strain of management practices based on principles of work intensification. Both groups of workers seldom reported their injuries. Las Vegas hotel room cleaners were discouraged from reporting injuries due to low rates of workers' compensation or complete rejection of claims when made. Both jobs were low-wage work, monotonous in nature, high in demand and low in worker control.

The relationship between work organization and MSDs has been examined by measuring job control and job demands [39, 40]. Psychological distress, strain, and muscular tension are viewed as resulting from problems of work organization, which in turn lead to muscular pain. Among women performing monotonous, rapid pace work, where they have little control over their work hours or conditions of employment, stress and a higher incidence of alcohol-related or gastro-intestinal illness, and hospitalization from heart attack have been revealed [41]. Comparing injury rates between sexes in different occupations, including service sectors, Smith and Mustard found that more than half of the difference in injury rates between women and men is due to the types of jobs they performed and industry type [42]. The findings revealed that women who worked in physically demanding manual jobs (like those in airport check-in work at semi-mechanized work stations) performed more repetitive tasks than males, and were more than twice as likely as men to suffer musculoskeletal injuries. Furthermore, women performing the same jobs with similar physical demands as male co-workers had injury rates as high or higher than the male co-workers, possibly because work equipment was designed to fit the average male and not suited for women in the same job. Women in the study had greater difficulty accessing workers' compensation (i.e., financial compensation) for musculoskeletal injuries, concordant with our study's findings of the non-reporting of MSDs among check-in workers [43].

The ten-year Framingham Heart Study showed that women with "high strain" jobs had nearly three times more chance of developing coronary heart disease than women workers in other jobs [44, 45]. And a 1993 Organisation for Economic Cooperation and Development (OECD) review of key studies on women's work and health revealed that women are more exposed to monotonous, repetitive work than men, and that their work content often can be characterized as high in demand and low in decision-making autonomy, the combination of risk factors producing the greatest job strain.

In many parts of the world, clerical work has been among the fastest growing occupational category in recent years. In the United States, approximately 80%

of clerical workers are women, and about one-third of all women workers are in clerical jobs. Other industrialized countries show similar trends. Offices, while seeming to be "safe" workplaces, often are ill designed and inappropriately constructed, with little attention paid to the health of workers [26, 46, 47]. Inappropriately placed lights, insufficient levels of lighting, too much glare, temperature extremes, poor ventilation, excessive noise, and poorly constructed seating are environmental factors that have been associated with adverse effects on health [48]. The use of computers has brought with it a range of health problems, most notably MSDs.

The conditions under which many *women* work, namely poorly paid, low-status jobs with a high level of demand and a low degree of control, have been identified as causes of negative stress in studies of *male* workers [49-51]. Evidence indicates the need to extrapolate the findings of negative stress and related adverse health outcomes to *women* performing high demand/low control jobs [38, 52]. Why should only men be so fortunate?

In the United States, women make up 46% of the work force, and 33% of those injured at work. Yet women account for 63% of repetitive motion injuries that result in lost work time (47,408 injuries out of 75,188). MSDs account for nearly half of all lost work time injuries and illnesses among women (see Table 1).

The higher occurrence of MSDs among women workers must be viewed within the context of the types of jobs that women usually end up performing in the labor market, characterized by low wage, low skill, high-demand/low-control, dominated by a high degree of hierarchic control, and highly stress-inducing. Airport check-in work typifies work of this general nature. A look at the three airports studied, the work performed by airport check-in workers there, the largely hidden exposures, and an analysis of the study findings reinforces what previous research results have informed us about women's work, particularly in service sector jobs.

Table 1. United States: Musculoskeletal Injuries among Women, 1997

Description of injury	Number of injuries causing lost work time to women
Carpal tunnel syndrome	20,584
Tendonitis	11,054
Injury due to repetitive motion	47,408
Injury due to repetitive typing or keyboard entry	10,131
Injury due to repetitive placing/grasping	14,950
Injury due to repetitive use of tools	5,117

Source: United States Bureau of Labor Statistics: Lost Work Time Injuries and Illnesses [8].

AIRPORT CHECK-IN WORKERS' TASKS
AND WORK STATIONS

Checking-in, the core task of the airport check-in worker, involves a series of various activities that require a variety of skills. Activities are cyclically repetitive. The typical sequence of activities to serve a single passenger follows:

- Welcoming a passenger
- Obtaining security information
- Receiving an air ticket
- Processing the ticket electronically
- Printing and issuing a boarding pass
- Informing a passenger about his/her seat on the plane
- Checking the weight of the baggage
- Printing a baggage tag
- Attaching the tag to the baggage
- Lifting and carrying baggage (often extremely bulky, heavy, and unwieldy) from the scale to the conveyor belt
- Activating the conveyor belt
- Making sure that the baggage moves along smoothly
- Picking up baggage when it falls off the conveyor and putting it back on the belt.

Additional regular tasks of check-in workers may include:

- Handling special baggage
- Unloading unclaimed baggage from baggage carousels and lining up bags on the floor to clear carousels for in-coming flights
- Gate duties, including lifting and carrying bags that cannot be hand carried onto airplanes by passengers and which have to be specially checked at the gate. Often the bags are heavy and check-in workers must carry them to the airplane tarmac.

Both sitting and standing present unique problems, even at work stations that allow workers to alternate between the postures for flexibility. For instance, considering that all work activities except lifting and carrying baggage could be done in a sitting position, Airport A check-in workers in Ontario, Canada nevertheless work mainly in a standing position. They stand because the few chairs they have (there are not enough for all workers) are broken, unstable or extremely uncomfortable and because there is no place for their legs when seated since the work station is extremely poorly designed, appearing more makeshift than anything else. If workers do sit, their legs hit against shelves under the counter (see Photo 1).

The impact of check-in work on workers can vary, depending on the design of the check-in counter and baggage-handling systems at any given airport. In general, the job of the check-in worker entails processing a passenger's ticket to generate a boarding pass and expediting a passenger's baggage for loading onto the aircraft. These two functions combine two considerably different kinds of demands, but each presents a risk of MSDs.

Factors that can influence a check-in workers' workload include the design of the computer work station, the baggage-handling system, the volume of passengers, and the regularity at which they arrive, the number of flights to check in per work shift, and the environmental conditions (particularly air quality, noise, and temperature) at each airport. It is unclear to what extent each of these factors impacts the workload for any individual worker and whether each factor has equal impact. However, the baggage-handling system and the design of the computer work station appear to have significant impact on workload, perhaps playing a more influential role in adverse health outcomes than the other factors.

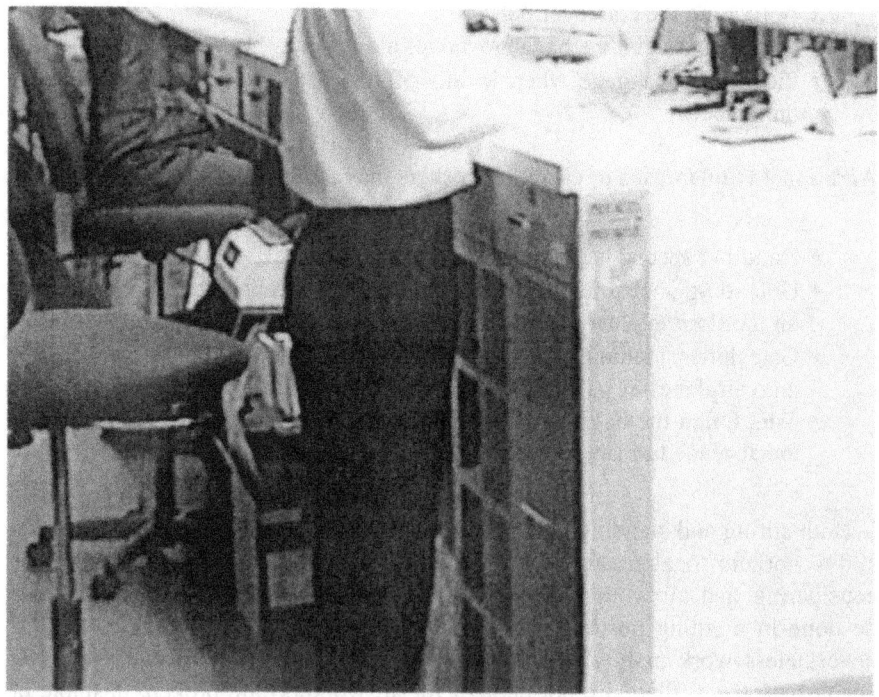

Photo 1.

It appears that airlines' quick turnaround policies also exercise a significant influence on workload. Work demand (exerted by the speed of work), the number of flights each day, treatment by management and pressure exerted on workers to ensure that passengers are checked-in quickly, combined with the emotional labor of maintaining a polite and friendly manner, contribute to a heavy daily workload. The work shift length plus number of bags handled per day (which fluctuates according to day and work shift) are additional indicators of workload.

Damaging workloads are indicated by the high rates of MSD-related symptoms revealed by workers at both semi-mechanized and fully mechanized check-ins, and among workers at very poorly laid out work stations (found at Airport A), as well as at better-designed work stations. Sitting for six to eight hours, and lifting or handling baggage, even where there is a fully mechanized system, plus twisting in an awkward posture to tag bags, in the absence of any training on proper lifting technique, appear to be hazardous combinations for workload levels and for worker health.

Workers and supervisors at the three airports studied commented on the absence of training on safe lifting technique and the need for the training.

> I never had any training on lifting. I don't even know if the employer has a policy on manual handling! I know we are allowed to lift 70 pounds, with the maximum allowed 100 pounds. (Worker, Airport A)

> I never had any real training on lifting, just I was told to bend my knees. I am careful when I have to lift baggage, to avoid back problems. I have a good idea just by looking at the size of a bag of the actual weight. (Worker, Airport B)

> I was supposed to attend a training seminar on safe lifting, but it was cancelled several times. I have no idea of any safe lifting limit. (Worker, Airport B)

> I never had any training on lifting, and we need this. Bags fall over behind my desk, and I have to ask passengers to help me lift them, because they are too heavy. (Worker, Airport C)

> I know there is a company policy of 70 pounds. I think I can safely lift up to 50 pounds. (Worker, Airport A)

> I've never had any training on lifting. If a big bag falls off, I'll ask someone else to pick it up. (Worker, Airport C)

> I never had any training on lifting, and I think there should be training on proper lifting. (Worker, Airport C)

There was a notable difference between Airport C in Switzerland and Airports A and B in Ontario and Québec. Two independent management companies operated at Airport C and check-in workers were employed by one of the two companies. One of the employers has a policy of not allowing check-in workers to perform solely check-in work, so that they performed multiple functions as part of their job, including a variety of passenger service duties consisting of working at the gate, assisting passengers in wheelchairs, helping elderly and disabled passengers, and responding to VIP passengers. While in principle, check-in work is meant to make up only part of the job, as it does for those Airport C workers, such flexibility often is not the case, and workers still performed check-in functions exclusively. Task variety is not built into the job of check-in workers employed by the other management company at the Swiss airport, where workers performed check-in work exclusively.

CHECK-IN WORK STATION LAYOUT

Detailed study of the work station layouts at all three study sites revealed a number of points in common. The work tables at check-in areas are at a fixed height and are not adjustable even though they have to be used by workers of different sizes. Computer keyboards and screens also are at fixed heights, even though they have to be used by workers of any size. Chairs may or may not be adjustable but have to be used in turn by workers of any height and size so they are not usually well adjusted. The work station may be designed for work in a standing or sitting position. Heat and humidity in summer and chilly drafts in winter aggravate the physical stress on workers' bodies at all three sites. A summary of the findings at each airport is presented below.

Airport A in Ontario, Canada

A small, regional airport with a semi-mechanized baggage system requiring workers to lift and carry every bag checked in.

The check-in work area is fixed, that is, non-adjustable, and is designed for sharing among a number of workers on various shifts. The work station is designed for work in a standing or sitting position at the discretion of the worker and consists of the following:

- A check-in counter typically occupied by a computer monitor
- Keyboard, printer, telephone, scale display, and a labelling/tagging device (see Photo 2)
- A tall sit/stand chair (not adjustable to the size of most workers)
- The baggage scale (see Photo 3)
- The baggage conveyor belt

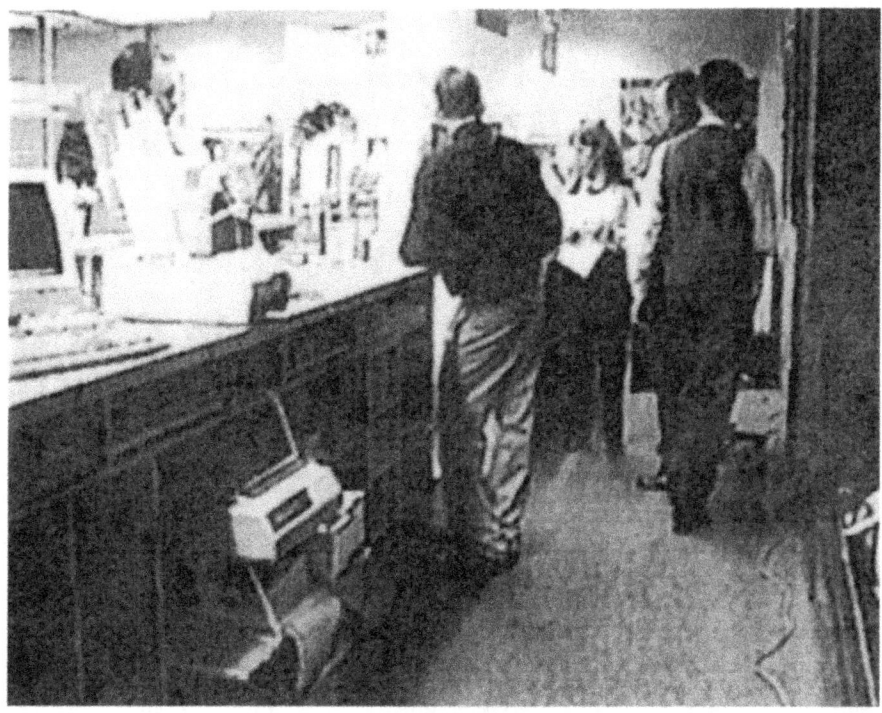

Photo 2.

Operating computers practically all day long at a work station lacking height adjustability and sufficient leg clearance contributes to a strained and uncomfortable posture, regardless of whether a check-in worker chooses to work standing or sitting. Manual lifting is a major source of discomfort and hazard for low back injury. Workers at Airport A bend low and reach forward to grasp and lift the baggage from where a passenger left it on the scale, most of the time using only one hand, then haul baggage to the conveyor behind their work station.

> We operate around one flight per hour, which is a rapid air schedule. We often have to pull bags off the in-coming baggage belt, which can be as much as a whole flight load of bags that have to be pulled off. I am the only one who will ever call for help or shut down the belt moving to get the bags off. Many of the bags are very heavy, plus they pull against a moving belt. (Worker, Airport A)

Repetitive lifting, twisting, and carrying, plus the heavy weights, put check-in workers at risk of musculoskeletal injuries, low back damage in particular. The

Photo 3.

environmental conditions at Airport A in Ontario aggravate the physical stress on the workers' bodies—temperature extremes, too much noise, and poor levels of lighting. Airport A check-in stations are near the entrance to the airport. There is one set of automatic doors between workers and the outside. Each time the doors open in winter, cold blasts of air hit the workers directly. In summer, the opposite problem occurs, with air conditioning inside the airport, and hot, humid air brought in from outside when the doors open.

In principle, work can be performed from a sitting or standing position, although workers spend nearly two-thirds of their work shift standing. Workers prefer to stand, even for extended periods, due to the complete lack of leg room under the check-in counter. The lack of leg room causes workers to hit their knees and shins against drawers or other hard surfaces when they work sitting down. Uncomfortable chairs are also obstacles to working from a seated position. The chairs are in a poor state of repair: some are broken, unstable, and unsafe. In addition, the unsuitably laid-out work surface, which necessitates over-reaching, excessive bending and stretching, also discourages workers from carrying out their jobs from a seated position. They are unable to get sufficiently close to the counter and computer screen due to the complete lack of space under the counter. The only means of reaching the computer in a relatively comfortable manner is to stand up and bend forward over the counter. Workers therefore generally prefer to stand during their work shift. Because of the conditions, more than 70% of workers at Airport A reported that they work in a constrained posture.

Standing work, typically linked with service-sector work, as at Airport A, has been associated with a variety of health risks, has a strong gender dimension, and has been little studied. Indicators show that standing work (where there is no possibility of sitting and where postures are relatively static) presents an important risk for workers in jobs typically performed by women. In a review of studies on the effects of work in a standing posture, Messing, Randoin, Tissot, Rail, and Fortin found that practically no study examined the degree of postural mobility among workers and no study considered workers' control over their work position [53]. Workers at Airport A, in theory only, had the possibility of working in a seated position. However, the reality of the unsuitable work stations and broken chairs made it virtually impossible to do so.

Airport B in Québec, Canada

Large international airport with fully mechanized baggage system where workers have the option to work either sitting or standing.

The check-in counter at Airport B appears similar to the one at Airport A with a few differences: the baggage-handling system has an additional conveyor belt connecting the baggage scale with the main conveyor, the computer work station has a height-adjustable keyboard tray and sufficient leg clearance under the counter (see Photo 4).

A number of features of the design and layout of the check-in counter at Airport B provide some degree of protection to check-in workers against aggressive passengers. These features include:

- The chest-high check-in counter
- The fairly narrow opening for the baggage scale
- The short stretch of the moving conveyor.

Check-in work stations at Airport B included a mechanical conveyor belt connecting the baggage weigh scale with the main conveyor. The conveyor system transporting baggage straight from the baggage weigh scale mechanized the baggage- handling system completely. This system was designed to speed up the check-in process. It has the further advantage of eliminating, in principle, the need for workers to lift and carry every bag checked-in. However, when the conveyor jammed, workers had to pull, tug, lift, and carry bags to un-jam it.

> We don't have to do much lifting of bags as such, but a lot of "un-jamming" of baggage. (Worker, Airport B)

Photo 4.

Similarly, when bags fell off the conveyor, workers picked them up and placed them back on the machine (see Photo 5). As well, workers often lifted and turned over bags as they were placed on the weigh scale to situate them in the correct position for tagging. Check-in workers also intervened manually in the case of special baggage checked in; they were the pieces of large size or awkward shape (see Photo 6).

The fully mechanized baggage-handling system appeared to decrease the workload, or the perception of workload, among check-in workers, yet despite the minimal need for manual lifting and carrying of baggage, Airport B check-in workers were still at risk for physical fatigue and postural discomfort. Tagging baggage, whether performed from a standing or sitting position, required considerable awkward body movements—bending, twisting, squatting, and reaching. In order to pick up the tag from the printer, a check-in worker had to twist the body (see Photo 7); then to attach the tag to the baggage, the worker had to bend over the baggage at around or sometimes below knee level—awkward postures repeated many times each day.

> We have to do a lot of twisting and stretching at the work station. (Worker, Airport B)

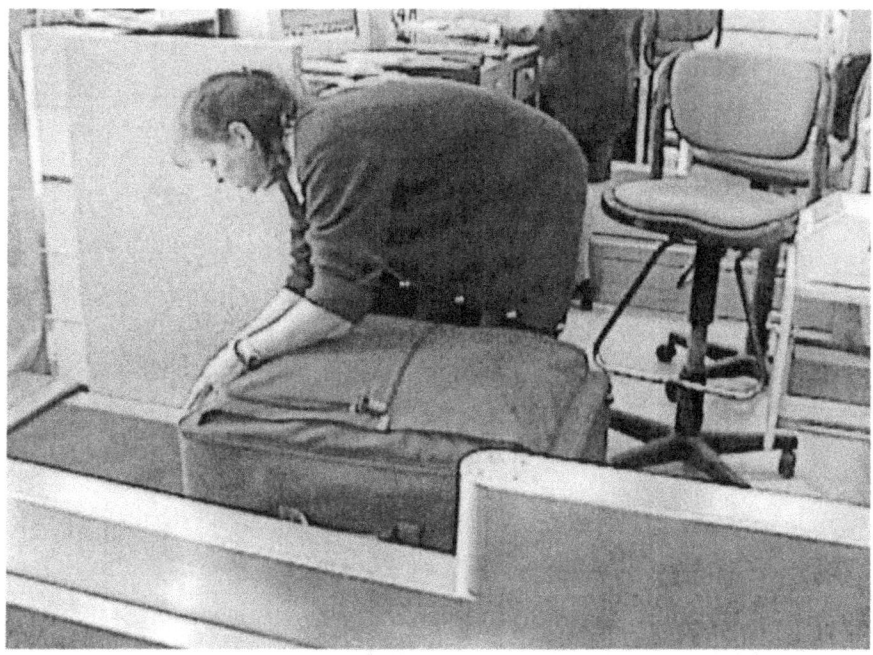

Photo 5.

> We have very little room to work. We do a lot of stretching, like reaching to get the passenger's ticket, and twisting to get the baggage tags. I remember one time after a particularly stressful day, I had neck pain for two days. (Worker, Airport B)

The computer work station had a height-adjustable keyboard tray and sufficient leg clearance under the counter (see Photo 8). Workers were provided with chairs, although the chairs were neither fully adjustable nor sufficiently ergonomic in design.

> We got new chairs with inclined seats, but no footrest. I have low back problems and do not like these new chairs; they are not comfortable for me. (Worker, Airport B)

Use of a computer through an entire shift at a work station lacking height adjustability can contribute to strained and uncomfortable body posture regardless of whether workers chose to work standing or sitting. Not all workers at the Québec airport appeared to know that the keyboard tray was height adjustable and therefore did not adjust it for their needs, a problem that could be rectified easily with information and training.

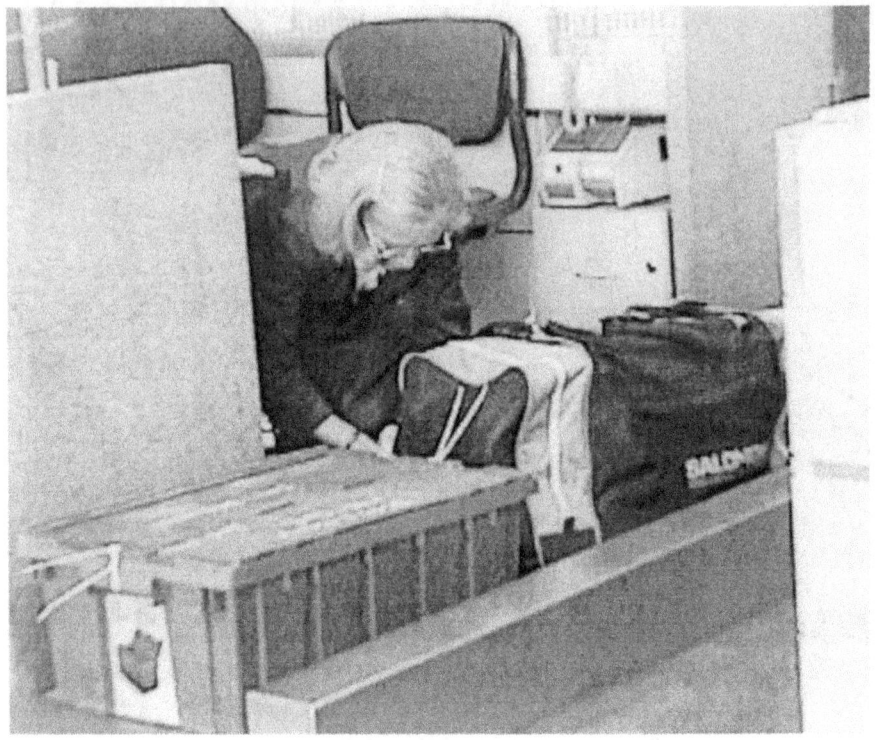

Photo 6.

Airport B workers reported that of the two check-in areas studied at Airport B—International and Domestic—check-in workers strongly favored the Domestic area. While differences in work station design between the two areas were relatively minor, workers favored the Domestic because there were worse environmental conditions, particularly thermal conditions and noise, and more difficult baggage handling at the International check-in. The baggage of international travellers often was oversized and heavier than domestic baggage. Larger "international" bags caused more frequent jams on the conveyor belts, thus forcing the check-in workers to intervene directly and physically to unplug the jams. Other challenges of the job, such as dealing with difficult or aggressive passengers and operating a computer terminal for long hours, remained similar to the challenges faced by check-in workers at the other study sites. The conditions caused fatigue in workers faster than at any other area at Airport B.

The International flight area is too warm. (Worker, Airport B)

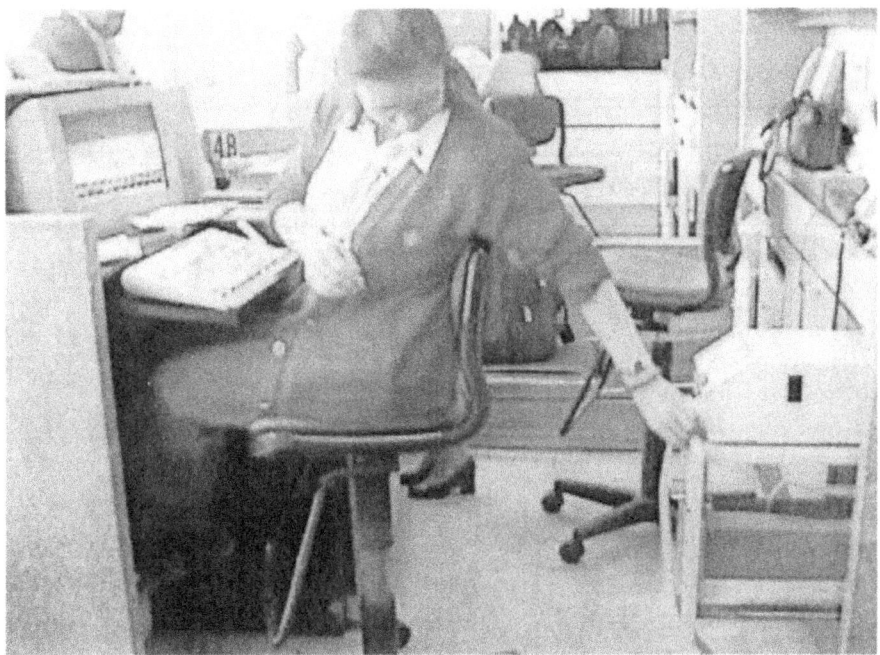

Photo 7.

The noise is worse in the International area and you get more tired there. (Worker, Airport B)

It is very warm in the International flight area (87 degrees) [Fahrenheit]. (Worker, Airport B)

The air quality in the airport is poor, there is a smell of diesel. Also, it is very dusty—to the point where some of the check-in agents got conjunctivitis. It is a very noisy environment—maybe acoustic tiles could improve the situation. (Worker, Airport B)

Heat and humidity seemed to build up faster at the International check-in area because of its low ceiling. Higher noise levels at the International area might have been due, in part, to the lower ceiling, which captured sound closer to the workers, as well as to the denser crowds associated with international travel.

Full mechanization of baggage handling combined with alternating work positions reduced postural discomfort at the work station.

Photo 8.

Airport C in Switzerland

A medium-sized international airport with a fully mechanized baggage system where workers sat through their entire shift.

Of the two ground-handling companies operating at Airport C, one offered the possibility of part-time work paying lower wages than the other employer and this company attracted students. Many workers indicated they did not intend to remain on the job for more than two years. In contrast, workers for the other employer were full-time and more likely to see their work as a career. They remained longer on the job. This company tried to vary check-in workers' tasks.

The check-in work station at the airport in Switzerland was designed for workers to perform their job functions from a sitting position during the entire work shift. Spacious L-shaped work surfaces (see Photo 9) allowed check-in workers to locate all of their work equipment (computer monitor and keyboard, baggage tag, and boarding pass printers) within easy reach (see Photo 10). Adequate counter space at arm level allowed workers room to perform their job functions, however, workers commented that the amount of space on the top of the counter was insufficient, particularly when checking in a group of passengers or a family at the same time. Airport C workers were able to view the baggage

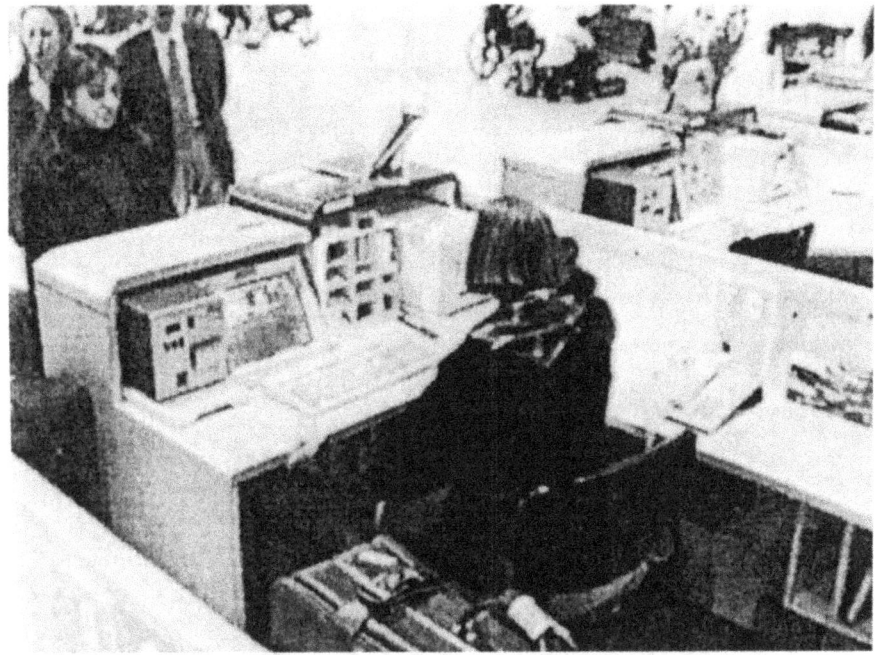

Photo 9.

scale display without twisting their head, neck, or any other part of their body. Semi-adjustable chairs allowed workers to sit relatively comfortably, particularly if they adjusted the chair for individual height. The baggage conveyor delivered baggage at the height and distance suitable for attaching tags without requiring workers to leave their chair.

In principle, Airport C workers did not have to lift bags that were checked in. As a result the check-in operation, with the exception of sporadic baggage jams, could be done from a sitting position, with check-in workers spending nearly 90% of their work shift in a sitting position. However, sitting and operating a computer for an entire work shift are significant hazards for postural discomfort and potential MSD. More than half of Airport C check-in workers felt that the job forced them to sit for too long. With only sporadic need to lift and carry baggage, and with all work functions within reach from a seated position, there was minimal objective need to adopt different work postures. The only opportunity to change from a sitting position occurred when the worker welcomed incoming passengers and while attaching tags to the baggage (see Photo 11).

The tops of the Swiss airport check-in counters met approximately chest level on an adult passenger. Check-in workers, from their seated position, reached up and forward to exchange documents with passengers. The counter height together

Photo 10.

with the worker's seated position placed workers below the height of standing passengers. Counters at Airports A and B were lower in height and passengers were able to lean forward across the top of the counter, coming relatively close to the worker's face while checking-in. Airport C's higher counter provided some degree of protection for workers against potentially aggressive passengers.

A fully mechanized baggage-handling system combined with a computer work station designed exclusively for work in a seated position might have reduced physical effort requirements, particularly compared with a semi-mechanized system. However, the lack of postural flexibility accorded by this design created conditions for postural strain due to prolonged sitting. A work station designed to allow both sitting and standing work, as well as easy adjustability for workers of different sizes, is preferable. Additionally, the fully mechanized system did not eliminate the need for workers to pull and push bags when they jammed the conveyor or to lift them off the floor when they fell off it.

Overall, the work stations at the airport in Switzerland appeared, at first glance, to be much better in design than those at the airports in Québec and Ontario. Workers at the Ontario airport had no leg room and were practically obliged to work standing, with only a semi-mechanized system at their disposal. Workers at the Québec airport could alternate their positions, but the placement of the

Photo 11.

tagging device required awkward twisting, and pushing/pulling bags was a common feature of the job, as at the Swiss airport.

WORK-RELATED INJURY REPORTS

The ground-handling companies at the airport in Switzerland and the airline employing check-in workers at the airport in Québec provided injury reports for check-in workers. Figure 1 summarizes injuries reported by the workers employed at Airports B and C for the period January 1998 to December 1999. In 1998 there were thirty non-disabling and twenty disabling injuries, which resulted in 280 lost work days. In 1999, there were twenty-six non-disabling injuries and fourteen disabling injuries, causing 181 lost work days. None of the employer-provided injury reports distinguished MSDs from other injuries, making it impossible to know whether MSDs caused any of the lost work time reports or whether MSDs resulted in workers' compensation claims. Statistics were not available for Airport A at all. Disabling injuries may have been highest at Airport C because of underreporting at Airports A and B.

The need to investigate the real costs to employers and workers caused by poorly designed work processes, work organization, and management practices

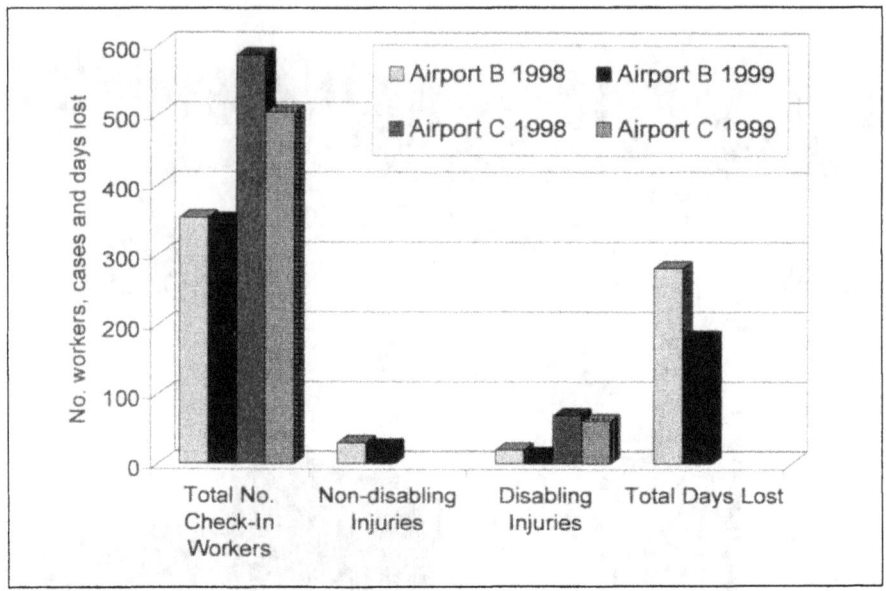

Figure 1. Check-in workers' injury report
(data provided by employers).

was punctuated by the significantly different lost work time injury reports
provided by the two employers of check-in workers at the airport in Switzerland,
which were presented in aggregate form. The difference in reporting appeared to
result, in part, from the fact that one employer hired very young, inexperienced
workers and paid lower wages than the other employer; it accepted a high
turnover of workers on the job and short employment duration. This policy meant
that workers were likely to leave the job before MSDs appeared.

WORKPLACE ENVIRONMENTAL FACTORS

Check-in workers at all study sites work on computers throughout their entire
work shift, a task that requires sufficient and appropriate lighting. Direct overhead lighting that reflects down on the computer screen makes it difficult for
workers to see information on the screen. Glare (from overhead lights or sun)
can cause computer workers to squint and contort the body in order to view the
screen. Twisting the body all day long to view a computer can rapidly lead
to musculoskeletal pain, particularly in the neck and shoulders. A significant
number of workers indicated that lighting was poor or inadequate in their work
environments.

The air is stuffy in the airport and the building is dirty. The lighting is really poor. Twice I had to go home with migraines from the lighting; it's fluorescent, compounded by the 12-hour shift. I hate the 12-hour shift, but the majority of the group prefer it. (Worker, Airport A)

Temperature fluctuations were a problem at Airport C in Switzerland. Significant numbers of workers at Airport C reported ambient temperatures that were either too hot or too cold. Temperature extremes contribute to general and prolonged discomfort, fatigue and muscular tension—particularly cold—as the muscles tighten while one attempts to stay warm. Temperature extremes appeared to result from poor temperature control in the check-in area, as well as from incoming draughts of hot air in summer and cold air in winter. Draughts were caused by the frequent opening and closing of the airport's automatic doors, which faced the check-in work stations. Uncontrolled variations in temperature caused discomfort to the workers. More than 90% of workers indicated that their work area was too hot some of the time, while 65% reported that it was sometimes too cold (Figure 2). Varying external climatic conditions, artificially regulated indoor temperature, and the combination of regular exposure to extremes of both heat and cold were all part of the daily work experience for many check-in workers. Temperature extremes and fluctuations can both cause or aggravate MSDs.

Other challenges of the job, such as dealing with difficult or aggressive passengers and operating a computer terminal for long hours, remained similar to the challenges faced by check-in workers in the other study site airports.

Noise has been shown to increase blood pressure and to cause negative mood [48, 54]. Although it is not a known cause of MSDs, noise, similar to temperature extremes and inadequate lighting, contributes to overall discomfort for check-in workers and causes stress. All factors considered together, the cumulative effects may well worsen MSDs, or worsen the perception of pain and suffering and create distress. Noise at the check-in work station was another reported disturbance for workers at the Swiss airport. Some 80% of workers said that their work area was too noisy and that noise interfered with their ability to perform the job properly and comfortably. Various noise sources were identified, including crowds in the airport, passengers checking in, passengers' family and friends accompanying them to the airport, the general ambient environment, airport public address systems, neighboring check-in work stations, the baggage conveyor, and the boarding pass printer. An additional and important source of constant background noise was the mechanized conveyor behind the check-in work station that took baggage to the airplane loading area. Noise was thought to be worse due to poor acoustics at the check-in areas. Constant noise on the job can add to fatigue and is an important work-related stressor.

Musculoskeletal problems resulted from a variety of causes among airport check-in workers. They are common, and they can lead to temporary or

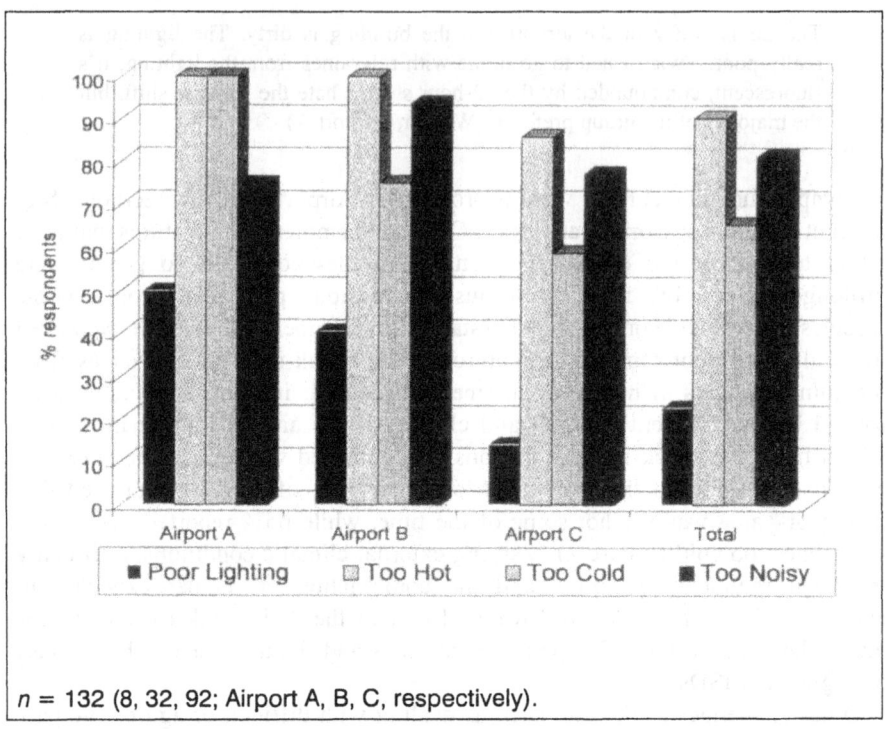

n = 132 (8, 32, 92; Airport A, B, C, respectively).

Figure 2. Environmental conditions
(% respondents selecting specified condition).

permanent disability. They are linked to lifting and carrying baggage, work station design, work organization, length of work shift at the conditions discovered, environmental conditions, and management practices. The findings were alarming.

MUSCULOSKELETAL DISORDERS IN AIRPORT CHECK-IN WORKERS

Neck and Shoulder Pain

I had tendonitis, bursitis, shoulder and neck pain. (Worker, Airport B)

More than 70% of the workers in the study indicated that neck pain affected work performance and nearly 16% had neck pain serious enough to take time off work. More than 70% experienced neck pain that interfered with their sleep, and

74% had neck pain that interfered with activities outside of work. Of the nearly 60% of workers who experienced neck pain, a greater proportion worked at Airport A. The findings indicated a major problem that should be addressed.

Neck pain was a direct outcome of lifting heavy bags (particularly at Airport A) and static posture or prolonged work at a computer station that lacked adjustability. That deficiency in the angle of the computer screen also may lead to painful neck strain, particularly among workers who wear corrective lenses. The findings indicate that the constant lifting of heavy baggage caused damage to workers' necks.

Shoulder pain also appeared to have a significant and negative impact on work life and non-work life: more than 70% of workers experienced shoulder pain to a degree that interfered with their work, more than 12% had to take time off work due to pain, more than 70% reported that pain interfered with non-work-related activities, and over 60% said that shoulder pain interfered with their sleep. Half of the workers who lost work time due to shoulder pain worked at Airport B.

Shoulder pain was a direct outcome of lifting and carrying baggage (particularly at Airport A), twisting and leaning to tag bags (particularly at Airports B and C), computer work performed at work stations that were not adjustable (all airports), as well as from lengthy shifts performing only check-in work.

Given that MSDs are cumulative over time, higher average age and mean length of employment at Airport B in Québec compared with Airports A and C in Ontario and Switzerland may explain greater numbers of the Québec airport workers losing work time due to shoulder pain. It is not, however, a consistent trend for the other forms of musculoskeletal pain. For example, nine workers at the Swiss airport, for example, lost work time due to lower back pain, compared with two workers at each of the other two airports, an outcome probably associated with prolonged sitting on the job, combined with awkward postures. Twisting to tag baggage, plus frequent pulling and pushing of bags, were also key sources of back strain.

Elbow, Wrist/Forearm, and Hand Pain

Sometimes I have pain in my upper body and arms . . . (Worker, Airport B)

Hand pain was reported less frequently than other types of pain, however, when it did occur, particularly at the Swiss airport, it was severe in consequence: more than 80% of Airport C workers with hand pain reported interference with job performance, over 80% reported disruption of non-work-related activities, and nearly 70% reported sleep loss due to hand pain. A few workers lost work time due to hand pain. Pain in the hand that caused interference with sleep, disruption to non-work activities, and interference with job performance could indicate tendonitis, or a severe, potentially debilitating nerve disorder, such as carpal tunnel syndrome. At the very least, such symptoms suggest the need for medical attention.

More workers reported wrist and forearm pain than elbow pain: 23 workers had wrist and forearm pain, and more than 20% of those lost work time due to the pain. Sixty percent said that it interfered with job performance; the same number said wrist/forearm pain interfered with non-work-related activities, and more than 50% said the pain disturbed their sleep. Wrist/forearm sufferers at the Swiss airport experienced pain to a degree that interfered with work performance, non-work-related activities, and sleep. This fact warranted attention since Airport C workers labored intensively at the computer in a fixed sitting posture throughout their entire shift. Wrist/forearm pain did not appear to be a widespread problem among workers at the Ontario airport where only one worker reported the problem.

Elbow pain seemed to be experienced less frequently than neck and shoulder pain, but among those workers experiencing it, the intensity of the pain appeared to be quite disruptive: fifteen workers experienced elbow pain; 80% said it interfered with work, more than 70% said it interfered with non-work activities, and nearly 70% said elbow pain interfered with sleep.

Elbow, wrist/forearm, and hand pains in check-in workers were likely caused by repetitive computer work at a keyboard surface that kept the arms elevated too high, from lifting heavy baggage and general baggage manipulation. The preponderance of suffering due to hand pain at the airport in Switzerland is most likely caused by the prolonged use of a computer keyboard placed on a table too high for most workers. The high table surface led to an overextension of the wrist, which can cause repetitive strain injuries of the wrist, arm, and shoulder. Computer users' wrists should be held in a straight position during keyboard use. The correct wrist position can be difficult, even impossible to maintain when the work station's height is not adjustable. The variation between sitting and standing work at the Québec airport may explain the difference in hand pain rates at these two airports. An additional key risk factor for hand, wrist, and arm injury is repeated lifting of heavy baggage. Work observation revealed that check-in workers at the Ontario airport lifted and carried bags with one hand most of the time, placing strain on the wrist, elbow, shoulder and arm.

Lower Back Pain

More than half of the check-in workers at all three airports lived with lower back pain and all workers at the Ontario airport suffered from this affliction. Lower back pain was less frequent at the Québec airport, probably due to the flexibility of working from both sitting and standing positions. Where lower back pain occurred among check-in workers, it was severe: over 60% of lower back pain sufferers said the pain interfered with job performance, 60% of lower back pain sufferers said the pain interfered with non-work activities, and nearly 80% from the Québec airport, more than 60% from the Ontario airport, and more than half from the Swiss airport said lower back pain interrupted their sleep. Lower back pain caused some check-in workers to lose work time, but not at a rate proportional to the number of workers living with pain. These findings indicated the existence of a severe problem.

Severe lower back pain at the airport in Switzerland was likely due to prolonged sitting in a static position, combined with, and probably aggravated by, awkward postures associated with baggage tagging. Lower back pain was common among workers who have to lift or pull heavy weights or who lift frequently, even if the loads are not heavy. It also was common among workers who sat for prolonged periods operating computers and among workers who adopted awkward postures, such as twisting and bending. The exposures were found at all three airports and explained the prevalence of lower back pain. Prolonged sitting and lifting caused more pressure on the lower back (lumbar region) than on the upper back.

Foot Pain

> Sometimes I have swollen feet. (Worker, Airport B)

Foot pain was experienced less often than other types of self-reported MSDs, but more than half of the check-in workers who reported foot pain claimed it interfered with their ability to perform the job or interfered with their activities outside work. The distribution of these two outcomes was nearly the same, experienced by workers at the Ontario and Québec airports. A significant number of workers from all three airports said their foot pain interfered with sleep. Where it was reported foot pain was more frequent and more severe among workers at the Ontario airport where workers stood on their feet performing check-in duties much more than workers at the other two airports.

Fatigue and pain in the feet, caused by standing for long periods, was made worse by regular lifting and carrying of heavy bags, which exerted significant pressure on the legs and feet. Prolonged periods of standing, particularly on hard surfaces, combined with regular lifting of bags were the most likely major factors explaining the foot pain in check-in workers, especially those at the Ontario airport. Where the chairs provided to workers were uncomfortable, or where there was a lack of leg room while sitting, workers were more likely to stand, contributing to foot pain. Adding a cushioned mat to the floor surface is a simple, low-cost means of increasing workers' comfort while standing.

Shoes constituted part of a check-in worker's uniform and, as such, had to conform to uniform regulations. Workers at the Ontario airport discussed their shoes with me. Airport A workers were provided an allowance of $70 Canadian per year to purchase shoes. The shoe allowance, however, was inadequate for the numbers of pairs of shoes workers typically wore out in a year: workers wore out an average of three pairs of shoes per year at work. Workers wore out their shoes quickly due to excessive standing on a hard surface, walking in the airport and on the tarmacs, and from lifting baggage, the weight of which placed wear and tear on the soles and sides of shoes. To stretch their shoe allowance, Airport A workers purchased low-cost shoes (in both price and relative quality) while keeping to the uniform requirements. When asked about the relation between shoe quality and back pain, workers appeared unaware of a connection; they said they had never previously thought about it. The workers had not considered asking for an increase in the shoe allowance, an issue for the health and safety committee. After our discussions, they decided to raise the issue with the committee.

Shoe quality and comfort bear importantly on the health of the musculoskeletal system, and are particularly important for workers who spend a great deal of time on their feet. Shoe quality is also important for back and leg support when lifting or handling weights, including baggage. It was likely that the quality of the shoes worn by check-in workers was a previously unrecognized factor contributing to lower back and foot pain.

Prior to this investigation, workers at the Ontario airport did not give high priority to purchasing shoes designed for support to the musculoskeletal system. This issue raised several points, such as why the workers did not seem to consider it their right to ask for a higher shoe allowance; whether lack of discussion of the allowance indicated that the health and safety committee needed to be more effective in its outreach to and general communication with workers; what were the mechanisms by which the committee identified issues for its discussions; and whether results were systematically communicated to the workers. Trade unions usually have numerous priority concerns to address at any given time, including political issues. Conflicting union priorities seemed to interfere with the ability to liaise sufficiently with workers on site at the airports.

Distribution of Pain and Impacts on Workers' Lives

It is not uncommon for check-in workers to suffer from MSD pain in more than one place on their body at the same time (Figure 3). A number of workers had pain in eight different points on their body simultaneously, indicating all over suffering. All-over pain was found at all three study sites, was not dependent on the type of baggage-handling system, and occurred among many workers. Some of the workers said they had to stop non work-related activities as a result of MSDs, ranging from four days to a year due to pain in one or several parts of the body. One worker had pain in seven different parts of the body, necessitating cessation of non-work activities for two months due to neck, shoulder, and elbow pain, and stoppage of activity for a year due to wrist, hand, upper back and lower back pain. These findings revealed a serious negative impact on workers' lives outside of work as a result of musculoskeletal pain, indicating work exposures

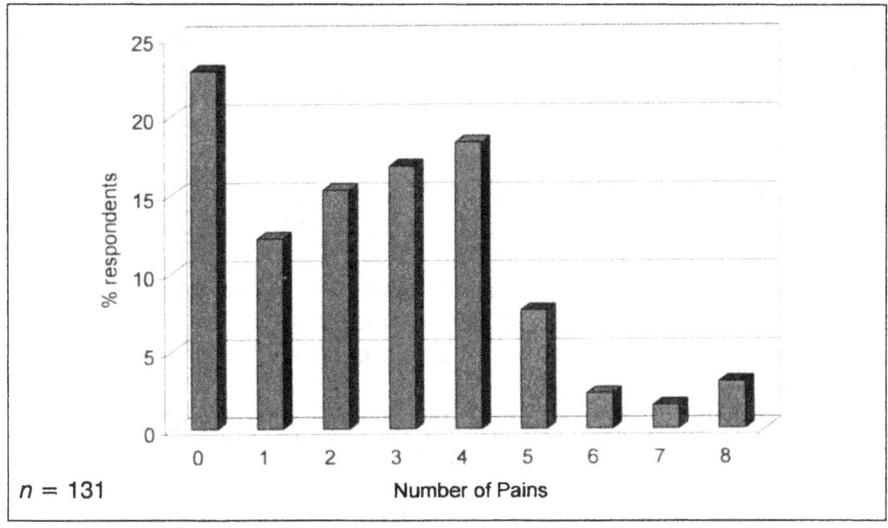

Figure 3. Total number of musculoskeletal pains reported
(% respondents reporting specified number of pains).

contributing to or as the direct cause of MSDs. Pain in multiple body sites is not, however, surprising, given that MSD-related pain from injury to one part of the body is seldom localized and frequently radiates to other parts of the body as well.

The discrepancy between the MSDs reported by workers compared with the injury reports provided by their employers showed that workers often do not report their MSDs to their employers. Official figures may, therefore, show zero non-disabling or disabling work-related injuries in a given year without reflecting the reality of workers' experiences. Yet when workers were asked about pain and suffering in an anonymous questionnaire, they revealed the existence of significant pain that interfered with work as well as sleep and life outside of work and was caused by their work conditions.

Reporting of symptoms was subjectively based on the feelings and perceptions of the workers. Whether and how much pain a check-in worker feels or reports may be dependent on the level of physical strain and psychological distress to which the worker is subject [22]. The significant number of check-in workers reporting living with MSD pain compared with the relatively low number of lost work days due to injury or pain indicated that many check-in workers considered musculoskeletal pain to be simply "part of the job." One might call this an "occupational culture of pain," wherein most of the workers lived in regular pain, but worked in spite of their suffering. The commonness in the experience of pain in the work group may serve to "normalize" to a certain degree how pain is perceived. Clearly, suffering from work-induced exposures should never be "normalized" or treated as acceptable.

Even when musculoskeletal pain interfered with the ability to perform the job or disrupted sleep or other life activities, check-in workers did not appear to lose much work time due to their pain. The contrast between the percentage of workers losing work time compared with the percentage reporting pain is significant: for example, nearly 80% of workers experiencing upper back pain reported that the pain interfered with their ability to work, while only 18% lost work time during the last year due to upper back pain. Similarly, 80% reported that elbow pain interfered with work, while only 20% took time off work due to elbow pain.

These findings can be interpreted in a number of ways:

a) Because pain was the norm at the check-in workplace, workers simply kept going to work despite their pain;

b) At a certain level, the pain was manageable. Those workers taking time off work may have suffered to a worse degree than other workers and/or may have a lower individual pain threshold (potentially a form of the "healthy worker effect");

c) Workers may have received subtle or overt messages from employers "encouraging" them not to take time off under workers' compensation due to cost implications for the employer;

d) Workers may not have taken time off under workers' compensation because they did not connect the particular pain to job exposures;

e) Check-in workers are a dedicated group and for the most part enjoy their jobs, which motivated them to go to work, even while living with pain.

Explanations c and d may be more significant, overriding explanations for the low rate of lost work time, given the high rate of MSDs. This supposition is based on knowledge of the workers' compensation systems in the two countries studied. Battling the "system" is not easy and is even more difficult when one is not in the best physical form (i.e., suffering from muscular pain). Workers' compensation systems in Switzerland and both provinces in Canada (Québec and Ontario) make it extremely difficult to obtain compensation for work-related MSDs. It is therefore difficult for workers to find a physician to support them, and difficult for them to get the "system" to support them thereafter. Thus, the path of least resistance would be to not register workers' compensation claims and not lose precious limited sick days to muscular pains, unless one is debilitated. Explanations a and e provide the requisite moral support that gives workers the courage to go to work each day in spite of pain.

How much discomfort or pain is acceptable for any worker to endure on the job? This is a question without answer, since the degree to which one can tolerate pain or discomfort varies from individual to individual. A more indicative question may be whether it is acceptable for any worker to suffer pain caused by job exposure, particularly when the exposures could have been prevented or eliminated? Often the nature and extent of suffering is not known, not revealed, until there is an investigation, or an effort is made to draw out, listen to, and address workers' concerns. Jobs are usually considered safe until they are studied and demonstrated as otherwise, despite the existence of any number of job hazards. Check-in workers work in an occupational culture where bodily suffering is the norm. Workers in this study did not seem to dwell on discussing their aches and pains; they appeared to adopt an attitude of "if everyone else can keep going, so can I." Workers seemed, more or less, to carry on with their jobs, despite pain and suffering (see Figure 3).

One might consider that the self-reporting of MSDs could lead to symptom exaggeration by respondents, consequently leading to an overestimation of prevalence. However, we found consistency in the responses about musculo-skeletal problems and symptoms among workers at the three airports, verified and reinforced by the work analysis and videotapes. It seemed, therefore, reasonable to treat the results with respect to subjective complaints as an accurate method for evaluating occupational health hazards and symptoms among check-in workers. The musculoskeletal disorders and injuries reported by workers in this study would appear to be largely (but not entirely) uncompensated injuries based on the low number of overall injury cases and days lost due to injuries evidenced in the employer-provided injury data, which are data for all injuries (see Figure 1). Considering the elevated level of self-reported suffering, the employer-provided data are presumed to significantly under-represent the true toll of MSD-related injuries and symptoms among check-in workers, which would make sense given the difficulty of obtaining workers' compensation for MSDs in Canada and Switzerland.

Naturally, workers may experience musculoskeletal pain that is induced by non-work-related activities, and which may worsen, even to the level of a disabling injury, due to work-related factors. The reverse can be equally true, where job factors can trigger a problem that could become aggravated by non-work-related activities. However, length of time spent on the job, cyclical repetition involved in the work, and regularity of daily exposures, including stress from a variety of causes, normally add up to more exposure time, exposure intensity, and in many cases more body load, than exposures caused by most off-the-job activities. This chicken-egg problem contributes to making the diagnosis (and compensation) of work-related MSDs somewhat tricky.

Without question, activities outside of work can cause or aggravate neck, shoulder, or back pain, or any MSD for that matter, depending on the activity undertaken. Given the significant prevalence of pain reported by respondents at all three airports, the exposures reported, and the amount of time spent on the job, it seemed reasonable to associate reported pains primarily with lifting and carrying baggage, awkward postures associated with job design, and muscular tension induced by psycho-social factors, including violence and emotional labor.

Sleep loss due to pain interferes with one's ability to perform optimally at work, reduces resistance to stressors on the job and outside work, and can catalyze a cycle of physician visits, treatments, medications, and time off work, all with cost implications to workers and their families, particularly when no causality to an on-the-job exposure has been established. In an occupation not previously investigated, it may be difficult for workers to establish a causal association between MSDs and work exposures for purposes of health insurance or workers' compensation, even if the worker herself first suspected the connection.

The employer-registered disabling and non-disabling work-related injuries related to check-in workers were far fewer than the level of self-reported MSDs. A significant number of workers in the sample population reported that they lived with any variety of neck, shoulder, elbow, wrist/forearm, hand, upper back, lower back, and foot pain. Workers at the Ontario airport said that all of the check-in workers they knew at Airport A lived with some form of musculoskeletal pain; they considered it "just part of the job." Concurring with these reports, our findings revealed that 100% of workers at Airport A lived with lower back pain. Such findings were worrisome, given that MSDs are cumulative with time and continued exposure. The relatively young average age of workers at Airport A (31 years) signaled the likelihood of a long work life living with chronic or sporadic MSDs and related pain and costs.

Punctuating this point, the impact of continued exposure over time suggested that as length of employment increases, so too does the ratio between the number of workers with and without symptoms. At least among workers at the two Canadian airports, more years on the check-in job, under then-current conditions, appeared to lead to a higher prevalence of workers with MSD-related symptoms. The same could not necessarily be said for the Swiss airport based on the evidence collected, since the average duration of employment was rather short. Some sort of breaking point might be surfacing; however, given the very different conditions between the two airports in Canada, any conclusion about this effect should be based on further investigation. These results do not mean that managers should adopt employment practices leading to short-term contracts. On the contrary, the job of a check-in worker can be modified so that it can be performed as a career, without adverse health outcomes.

Lifting and Handling Baggage

All of the check-in workers at Airport A in Ontario, where the baggage check-in system is semi-mechanized and obliged workers to lift and carry every bag checked in, reported that they manually lifted baggage as part of their job, and did so often. And even though fully mechanized baggage check-in systems existed at Airports B and C in Québec and Switzerland, a significant percentage of workers still lifted or handled baggage manually. While more workers at these two airports reported that they only performed such manual handling sometimes, there nonetheless were workers reporting that they manually handled baggage often.

The average number of bags handled per day (Table 2) varied rather significantly between the three study sites. The highest maximum number of bags reported handled (650 bags per day) and the highest average number of bags handled per day (418) occurred at the airports in Switzerland and Ontario respectively, the latter where the check-in system was only semi-mechanized. Even with the fully mechanized check-in systems at the airports in Québec and Switzerland, workers reported handling on average from 97 to 135 bags per day. This meant that at Airport A in Ontario, workers lifted and carried an average of 13,794 kilograms per day (based on the 33 kg maximum allowance per bag in North America). No wonder they all had back pain.

Pain and MSDs among check-in workers at the Québec and Swiss airports appeared to result, at least in large part, from prolonged sitting or standing and awkward body movements, rather than from constant manual lifting of passengers' baggage, except for lifting and carrying fallen bags and un-jamming the conveyor.

Workers manually handled baggage at all three airports, albeit to different degrees: all Airport A workers manually lifted baggage often (Figure 4). Fully mechanized baggage check-in systems notwithstanding, nearly 80% of Airport B workers and over 60% of Airport C workers lifted baggage manually (Figure 5).

Handling of baggage took place mainly at three locations: at the initial check-in, at the gate, and at the special baggage area (Figure 6). Special baggage is baggage that is overweight, oversized, unwieldy or awkward in shape. Skis, bicycles, and pets of any size in cages were included in the category of special baggage. Special baggage handling presented an additional burden for workers and was associated with awkward movements in addition to those at the standard check-in. With no training in safe lifting technique or mechanical lifting aids, workers lifted items of extreme weights and of any odd size at the special baggage handling area. Workers bent from the waist, twisted and reached to wrap, box, lift, and carry special baggage. Check-in workers' jobs included lifting, carrying, and generally handling special baggage, regardless of size or weight.

Table 2. Number of Bags Handled Per Day, Per Person

	Minimum	Average	Maximum
Airport A	160	418	600
Airport B	2	135	500
Airport C	0	97	650

n = 91 (8, 20, 63; Airport A, B, C, respectively).

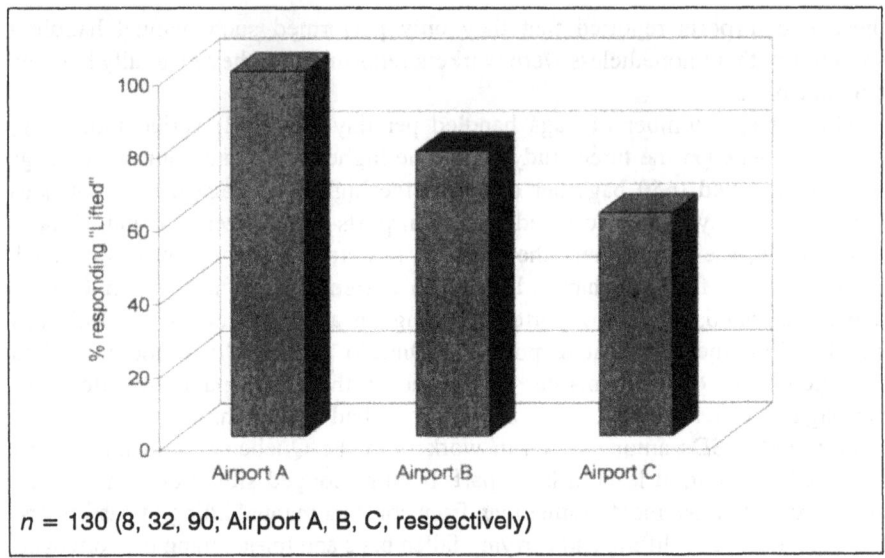

Figure 4. Whether baggage lifted or generally handled.

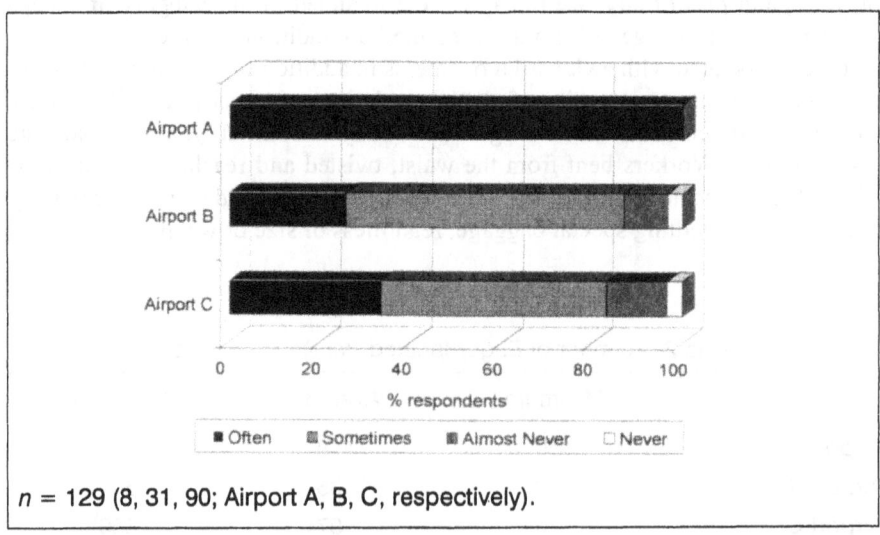

Figure 5. Frequency of handling baggage.

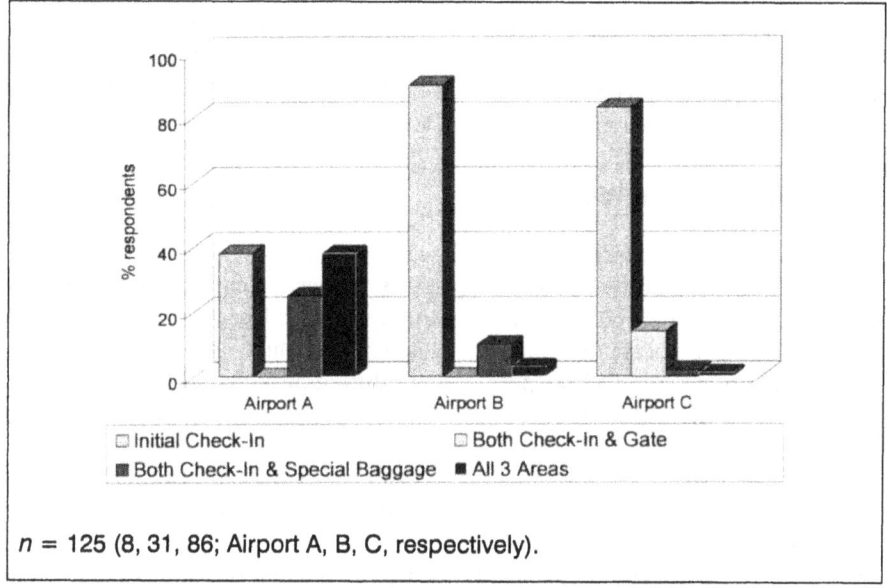

Figure 6. Areas where baggage is handled
(% respondents dealing with specified areas).

Workers at all three airports handled baggage at the initial check-in, where the greatest burden of work was performed. At Airport B in Québec, workers also handled special baggage. At Airport C in Switzerland, workers handled baggage at both check-in and gate. Gate duties involved carrying baggage to the airplane directly, including lifting and carrying bags up and down stairs, while baggage weights were *not* specified on the bags workers had to carry.

No system is foolproof. Even where there was a fully mechanized baggage--handling system, workers still ended up lifting and carrying bags some of the time.

> Last year I had one month off due to an accident at the check-in. I had a medical certificate and got paid. I slipped and fell off the step behind the desk and tore ankle ligaments. I don't want to hurt my back, so I only lift bags when they fall off, but bags fall off often. If a big one falls, I will call for help. (Worker, Airport C)

Lifting such heavy loads sometimes exceeded existing recommendations for safe lifting limits, such as those set out in the NIOSH guide for manual lifting [55]. Failing to follow safe lifting limits though was not without consequence: the high number of workers who lived with back pain was a result of unsafe conditions for manual lifting.

Quick turnaround policies appeared to be the root of the problem of work speed and work intensification, which directly affected the handling of baggage by check-in workers. The pressure, workload, and speed demanded of check-in workers contributed to negative physical and psychological health outcomes.

Computer Use and Work Posture

Fully computerized processing of passenger tickets involves health hazards similar to those faced by computer operators in clerical environments, notably fixed and constrained postures. An externally imposed pace of work, such as a high volume of passengers, the management of which is dictated by quick turnaround policies, and a high pace of work during peak periods, also presented potential risks for MSDs in check-in workers, similar to computer operators in clerical environments.

With work shifts ranging in length from six-and-a-half to twelve hours (Figure 7), check-in workers spent most of their working day in front of a computer. Nearly 100% of check-in workers spent almost their entire shift in front of a computer (Figure 8). Body load and potential damage resulted from static posture maintained over prolonged periods of time (Figures 9 and 10 and Photo 8) and conditions found at all three airports. They included work at poorly designed or non-adjustable work stations (including computer height and angle, table and chair), use of chairs that did not adequately support the back and legs,

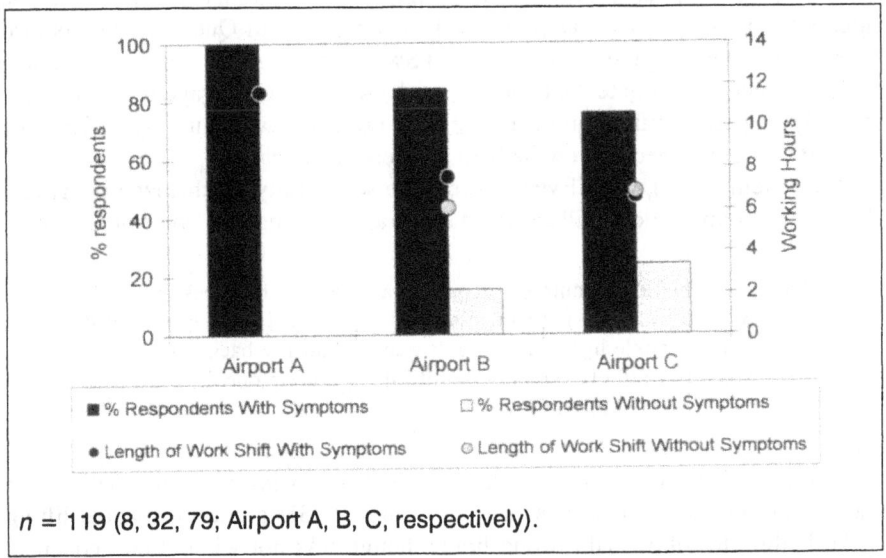

n = 119 (8, 32, 79; Airport A, B, C, respectively).

Figure 7. Length of work shift.

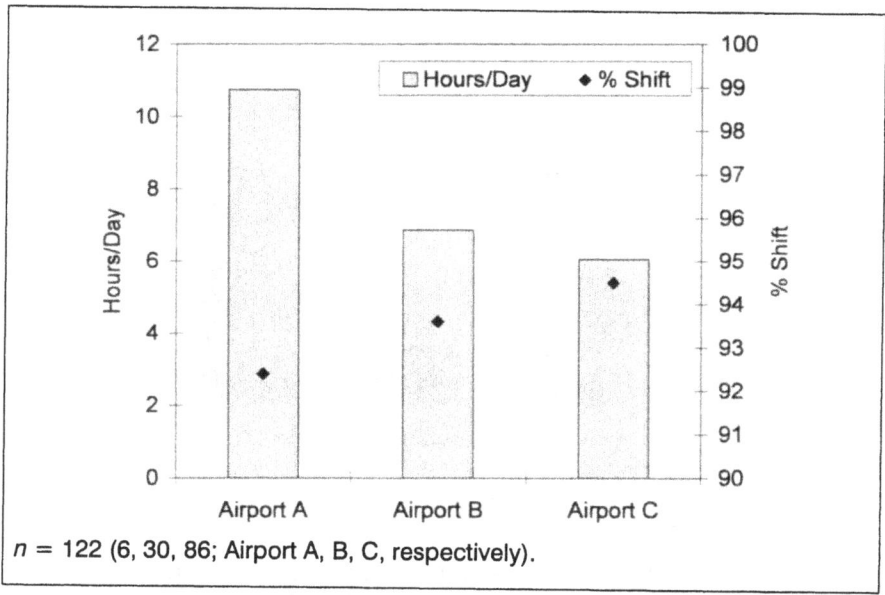

Figure 8. Computer use (number of hours per day and percent shift spent on computer).

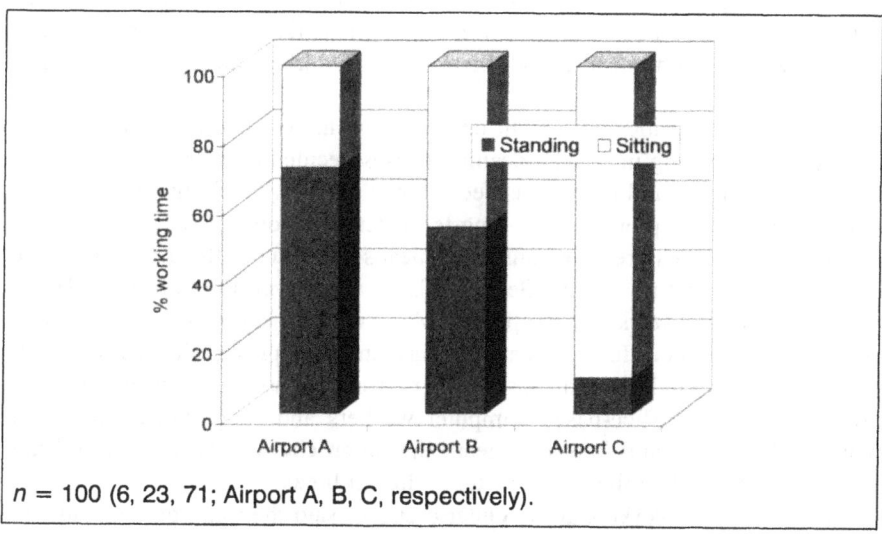

Figure 9. Work posture (% work time spent in specified posture).

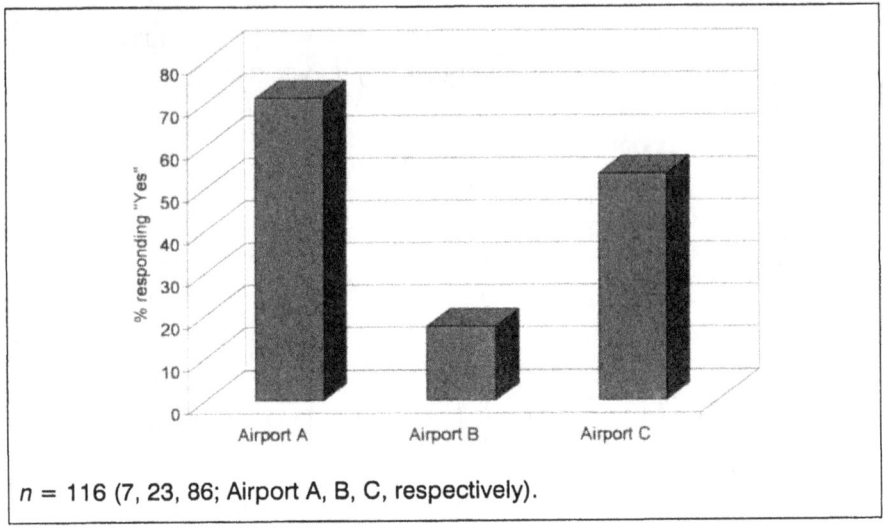

$n = 116$ (7, 23, 86; Airport A, B, C, respectively).

Figure 10. Spending too much time in one work posture
(% workers responding "yes").

repetitive hand movements, glare from overhead lights or direct sun, lack of regular breaks, and pressure to work at a certain speed. A worker at Airport B commented, "The fact that the VDU monitors cannot be adjusted is a problem. I get frequent eye strain." Adjustability was particularly important for check-in workers who worked on shifts and thereby shared work stations with workers of different sizes.

Nearly seven decades ago, Firth observed the importance of regular pauses at naturally occurring intervals, meaning pauses decided by workers themselves [56]. Preventing natural breaks has been shown to result in a long-term accumulation of strain symptoms and a long-term deterioration in work capabilities. Regular work breaks are a demonstrated means of reducing the effects of static posture and preventing MSDs. Benefits also are achieved through job enlargement, decided with worker participation, where workers can perform a variety of tasks. To reduce the effects of static posture and help prevent MSDs, check-in workers, toiling regularly and consistently in front of a computer, require regular breaks, similar to office-based computer workers and supermarket check-out workers. Workload and intensity meant that often check-in workers could not even stop to go to the toilet let alone take a longer break.

The majority of workers at the Ontario airport said their computers could be adjusted, and roughly half of the Québec and Swiss airport workers said they could adjust them, while half said they could not (Figure 11). Yet work station analysis revealed that, in fact, none of the computers at any of the airports were

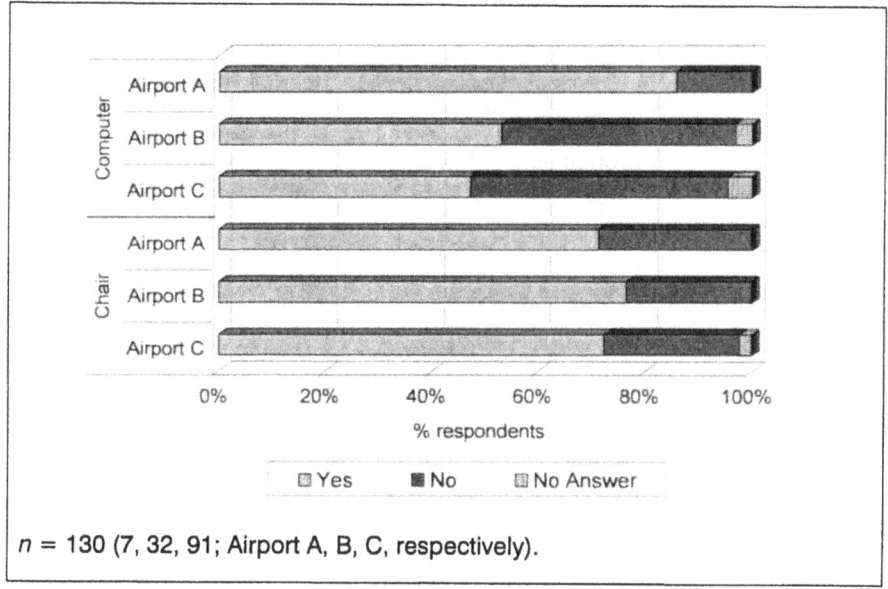

n = 130 (7, 32, 91; Airport A, B, C, respectively).

Figure 11. Can (Yes) or cannot (No) adjust computer and chair.

adjustable in height and only some were adjustable in tilt. Computers at the Ontario airport were adjustable neither in angle nor in height.

Over 75% of workers at all three airports said they could adjust their work station chair, while roughly one-third at each airport said they could not adjust them (Figure 11). Yet here too, as with the reported computer adjustability, work observation revealed that chairs at Airport A in Ontario were not adjustable, that some chairs at Airports B and C were adjustable, while others were only partially adjustable (such as in seat height, but not back rest angle or height).

Did adjusting the computer or chair make a difference in whether one had pain from MSDs? Not particularly. Pain appeared to be omnipresent regardless. Forty percent of workers who believed they could adjust their computer reported living with MSD-related pain nonetheless, while 35% who could not adjust their computer said they lived with MSD-related pain. Similarly, over half of the workers who said they could adjust their chair still experienced pain from MSDs, while over 20% of workers could not adjust their chair and lived with pain. Workers lived with MSD-related pain whether or not they could adjust their computer or chair. There was no statistical difference, although adjusting the chair did appear to make some difference (Tables 3 and 4).

Check-in workers at the Québec airport took advantage of the existence of a chair to sit for half of their work shift (Figure 9), which was recommended to decrease body load over the work shift, to increase blood circulation in the body,

Table 3. Can Adjust Computer

130 responses	With symptoms		Without symptoms	
	No. of respondents	Percent	No. of respondents	Percent
Yes	52	40.0	14	10.8
No	46	35.4	13	10.0
No answer	2	1.5	3	2.3

Table 4. Can Adjust Chair

128 responses	With symptoms		Without symptoms	
	No. of respondents	Percent	No. of respondents	Percent
Yes	72	56.3	22	17.2
No	27	21.1	5	3.9
No answer	0	0.0	2	1.6

and generally increase comfort level. Workers at the Ontario airport, where chairs were not available or broken when they were available, may have perceived that they sat down for part of their work shift, but work observation revealed that their "sitting" was actually leaning against the counter or sitting in the back room and not actually at the check-in work station.

Airport check-in workers should be provided with fully adjustable chairs: adjustable in seat height and tilt, back rest height and angle, with a five-wheel base for stability and ease of movement, and a fabric-covered seat pan to prevent sliding. Training should be offered to demonstrate how to adjust the furniture and how to determine appropriate chair height for each individual. Chairs should be adjustable easily from a sitting position.

Back pain, neck pain, shoulder pain, elbow pain, wrist, hand, and foot pain— these pains do not appear to discriminate between the sexes. Figure 12 provides an overview of the distribution of these symptoms by sex. Overall, women check-in workers seem to suffer shoulder pain much more than men, most likely from lifting and carrying baggage and from static and awkward postures at the work station. The significant sex-based difference in shoulder pain may reflect simple strength differences between male and female workers, a difference particularly apparent in the manual handling of weights, such as baggage. Men

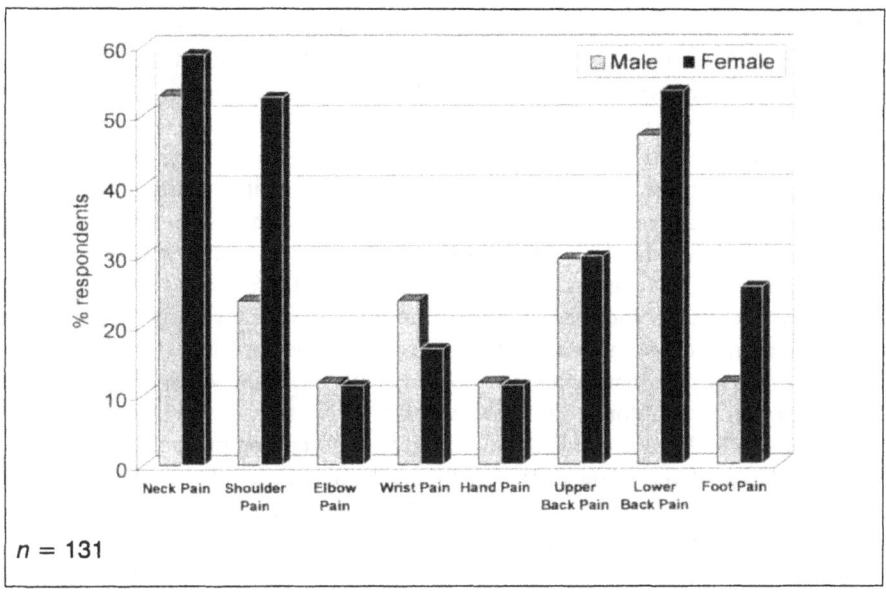

Figure 12. Distribution of musculoskeletal symptoms
(% workers reporting specified type of pain).

appear to suffer wrist pain more than women, possibly due to ill-adjusted computer keyboard and work station height. And women suffer from foot pain much more than men, which most likely has something to do with the fact that men can wear solid, closed-front, sturdy, arch-supporting shoes as part of their uniform, while women have to wear pretty little shoes with heels as part of their uniform. Such shoes are generally not made for heavy work, carrying loads, excessive walking on tarmacs, standing a great deal on hard surfaces and usually do not provide good arch and back support. Standing on hard surfaces for prolonged periods certainly explains a great deal of foot pain experienced among women workers [57].

In an attempt to investigate whether check-in workers' MSDs also may be associated with the length of the work shift, the number of workers having pain and those having none was cross-tabulated with the length of their respective work shifts. The findings showed that at Airport A, where the work shift was between 11½ and 12 hours long, all workers had MSD pain (Figure 7). At Airport B in Québec, significantly more workers on a 7½-hour shift had MSD-related pain compared with those on a roughly 6-hour shift. And at Airport C in Switzerland, where both groups of workers worked an approximately 7½-hour shift, three times more workers had MSD pain than workers having no symptoms.

The findings revealed that as length of employment increased, so did the ratio between the number of workers with and without MSD symptoms. In other words, under the studied conditions at the three airports, more time spent on the job appeared to lead to a higher prevalence of workers with MSD-related symptoms (Figure 13). It appeared that the symptomatic category of workers began to increase significantly after three years on the job under present conditions, although MSDs were shown to exist even among workers employed as little as six months or less. The U-shaped curve shown in Figure 13 is frequent when examining symptoms over time on the job. It shows that early on in the job there is a self-selection of workers who cannot stand the conditions and who leave. Later, among workers remaining on the job, some will adjust themselves to the conditions, while others will learn means of coping with the job. But with time, shown by the spike in those expressing symptoms, the accumulation of wear and tear reveals itself among those who have tried to adjust or tried to learn coping mechanisms.

These findings may indicate a relative tolerable limit of body load for a check-in worker, after which point musculoskeletal pain begins to appear. It seemed that approximately six hours of check-in work might be the "safe" limit at a work station if the worker can sit at will and has a fully mechanized baggage system. One-and-a-half hours more work time under the same conditions resulted in a significant increase in the number of workers who experienced MSD pain.

For workers lifting and carrying baggage all day long, there was likely no "safe" exposure limit under such conditions. However, further investigation of

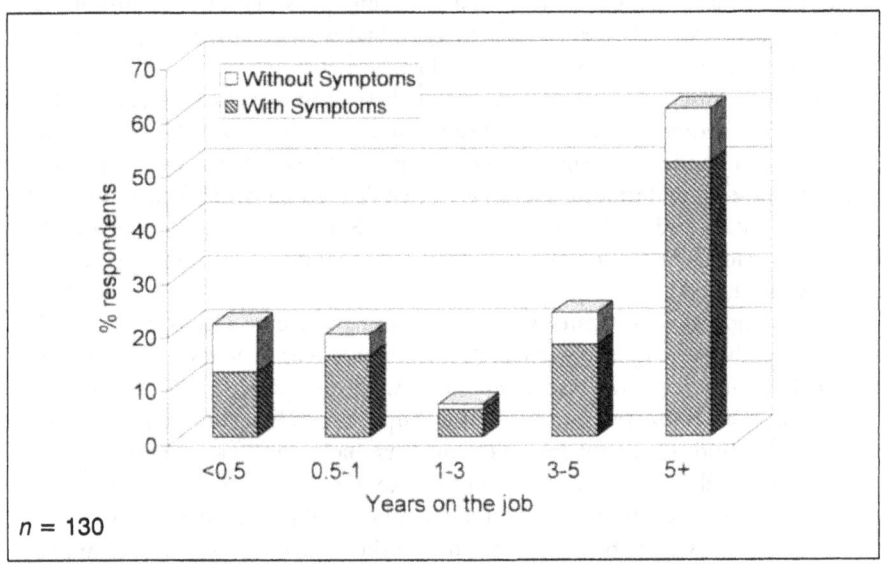

Figure 13. Distribution of symptoms by years on the job.

workers at semi-mechanized systems working various shift lengths is necessary to draw more firm conclusions. What was evident, however, was that workers employed under conditions similar to those found at Airport A were extremely likely, if not certain, to develop MSDs. For check-in workers operating under conditions similar to those at Airport C (fully mechanized baggage system, sitting down throughout the entire shift), MSDs there too were likely to develop, largely due to excessive static posture combined with various awkward movements, even when the work shift was less than seven hours.

Rather than "tolerable limit of body load," perhaps a "breaking point" was, in fact, revealed. The cumulative effect of the various contributing factors over a given period of time, with or without the moderating effect of working in an airport, may cause a breaking point in a worker's physiology. If this is the case, the physical characteristics of a worker, notably relative strength, would be rather irrelevant, and most workers would end up with some sort of health problem, given sufficient exposure level and frequency. This notion does not mean that check-in workers should be limited to three years on the job, in the expectation that MSDs would develop thereafter. Full mechanization of baggage check-in systems alone was not sufficient to eliminate MSDs from check-in workers, as demonstrated by significant rates of MSDs among workers at the Québec and Swiss airports. The findings indicated the need to re-examine the entire job design, organization of work, and management practices, and to put in place systems for the prevention of harm to workers from psycho-social stressors, including violence.

MSDs in airport check-in workers were associated with physical factors, including the manner in which work is performed based on conceptualization of the job tasks, level of mechanization for baggage check-in, work content, physical postures adopted by workers as determined by job tasks and work station design, pace of work, the amount of time a worker is exposed to the conditions of the job, exposure to various workplace environmental factors (such as extreme cold or heat, noise, insufficient lighting, or glare that inhibited viewing the computer screen), organization of work, and management practices. While all of the identified factors were important, the level of mechanization of the baggage check-in system appeared to have a stronger influence on the outcome of MSDs than other factors in airports where the system was only semi-mechanized and thus required workers to lift and carry every bag checked in. However, in the final analysis, it was indeed management practices that determined all of the above factors to one degree or another.

SUMMING UP MSDs IN CHECK-IN WORKERS

MSDs were prevalent and severe among airport check-in workers and may lead to temporary or permanent disability. The three most important MSDs identified in this study were shoulder, neck, and lower back pain. More than half the workers in the study experienced neck pain, half experienced shoulder pain, and one out of two lived with lower back pain. Significant numbers of workers lost work time due to neck pain, lost time at their professional activity from shoulder disorders, and had missed work due to lower back pain. A substantial number of check-in workers performed their job functions despite significant neck, shoulder, and/or back pain (defined as a level that interfered with the ability to perform one's job or that interfered with sleep). They co-existed with neck pain that interfered with their ability to work, and shoulder pain and/or lower back pain that interfered with work performance.

Well over half of the workers had lower back pain that interfered with sleep. More than 70% reported living with neck pain severe enough that it interrupted sleep, while an equally significant number lost sleep due to shoulder pain. These findings showed that airport check-in work had clear hazards associated with the job, an occupation likely to cause severe MSDs.

Important differences between the MSDs at Airports A and B in Canada, compared with Airport C in Switzerland, may be explained by the shorter duration of employment of many workers at Airport C. Higher worker turnover was revealed at Airport C where many more workers spent fewer years on the job than workers at the other two airports (average duration of employment at Airport A was nearly ten years; it was nine years at Airport B, while workers at Airport C remained on the job an average of four years). The shorter duration of employment at Airport C was not surprising, given the existence of two employers there, one of whom paid lower wages, sought younger, less experienced workers, and allowed worse work conditions. More respondents employed by this employer may have returned questionnaires than did workers for the other employer.

No obvious, significant differences of levels of MSDs appeared between men and women. Both sexes were confronted with the same exposures while performing check-in work. The strain of lifting and carrying baggage, awkward postures, and excessive sitting appeared to know no sex bias, although proportionately more women were affected due to their greater numbers in the occupation. Where work entailed static posture, both male and female workers were at risk of MSDs. Where work stations were not adjustable and prolonged computer work was performed, exposures were the same for both male and female workers with the same potential for MSDs as an outcome. However, since men were proportionately less represented in the sample, further investigation, with over-sampling for male check-in workers, is needed to draw firm sex- and gender-based conclusions about exposures and outcomes.

Lifting, carrying, or generally handling baggage, appeared to be sufficient to cause injury and MSD-related pain, even if the check-in system was fully mechanized. Baggage tagging also appeared to be a sufficient cause of MSD pain, since the task involved awkward twisting and bending at all airports observed.

Indeed, the study data indicated that MSDs were much more prevalent, and far more severe, among check-in workers than what was revealed in the official work-related injury data obtained from available employers' records. It was cause for concern when workers reported suffering from aches and pains in multiple body parts at the same time, particularly when four workers reported living with pain in eight different points on their body at once. This level of pain constituted all-over suffering. Experiencing multiple points of pain was not an isolated event either; 24 workers lived with pain in four different body points at once. In fact, only around 20% of check-in workers reported *not* having any musculoskeletal pain, while nearly 80% lived with pain from MSDs in one to eight different sites on their bodies.

MSDs in this worker population caused significant disruption to activities outside of work. Significant rates of interference with sleep due to MSD pain and negative impacts on job performance were revealed. Work efficiency and performance level suffered when abilities were hampered due to pain and loss of full range of motion, often resulting from MSDs. Sleep loss, moreover, can increase the risk of a work-related accident. While the costs to employers from lost work time due to sickness absence and decreased productivity are direct and measurable, indirect costs to both workers and employers are more difficult to measure. However, there are costs associated with reduced efficiency occurring from sleep deprivation or sleep interference due to pain, as well as costs associated with decreased productivity and worker turnover.

The fact that many MSDs were not registered with respective employers indicated a need for awareness-raising and education. The discrepancy between the self-reported levels of MSDs and the officially registered injuries indicated a need for trade unions and pro-active management to conduct awareness-raising among workers and managers to highlight the importance of registering workers' work-related health problems. Workers should be encouraged to report MSDs to their employers and the union, whether the MSDs are disabling or not and whether lost work time results or not. This strategy is a way to establish a health registry that management can use to justify any costs associated with improvements and to record gains in workers' health over time.

Many managers prefer to ignore workers' health problems, including work-related stress. However, there also are concerned managers who recognize the benefits of protecting and promoting workers' health, which leads to motivation, productivity, and profit. Indeed, truly pro-active managers seek out such information rather than passively wait for it. Eliminating the problem, or at minimum addressing it openly, can catalyze positive change for both workers and managers, leading to improvements in productivity and increasing well-being among workers. Such an approach is a key element of a pro-active management policy. Work station re-design, increased flexibility in work tasks, adjustable work station furniture, with workers' involvement in the decision-making process, therefore constitute minimum solutions for reducing MSDs among check-in workers. The results from this study provide evidence for managers to justify work station design alterations that allow workers sit/stand flexibility. The findings confirm as well that flexibility in work position reduces body load.

It is difficult for unions or management to determine where problems are developing unless workers report symptoms and health problems. Workers may not report MSDs for fear of losing their job. A current trend in the airline industry is to encourage senior workers to leave their jobs and to replace them with younger, inexperienced workers hired for lower wages. This labor market and employment insecurity may be a key motivation for senior-level workers to not report their MSDs. But younger workers may turn over faster than workers with seniority, which implies increased costs to management for new worker training, reflecting a short-term management vision at best. Interviews with workers at Airport C in Switzerland revealed that younger workers generally do not expect to stay at the check-in job for more than a few years, due to low pay, poor work conditions, and work that no longer provided reason to feel motivated.

Additionally, when younger workers leave the job after only a few years, it is difficult, if not impossible, to link check-in work exposures with any serious MSDs that appear later unless problems were registered during the period of employment. Management may find this scenario appealing, since it effectively relieves them of responsibility and cost. As workers and society (through health insurance costs) end up bearing the costs associated with work-related adverse health outcomes, unions may want to establish workplace registries to track exposures and outcomes.

From a management point of view, a long-term cost-benefit vision that considers factors affecting workers' long-term commitment to productivity, quality of service, commitment to job and employer, and well-being, is more beneficial to both workers and employers than a short-term vision.

Fully computerized processing of passenger tickets with work performed while only sitting or only standing—that is, in a fixed posture—involves health hazards similar to those faced by computer clerical workers and supermarket check-out workers. Where baggage check-in systems are manual or only semi-mechanized, the physical work required of check-in workers involves health hazards similar to those confronted by airport baggage handlers, a predominantly male population whose conditions of work have been associated with high rates of back injury, disability, high rates of lost work time, and high levels of workers' compensation. Work-related factors that present the greatest risk for discomfort and eventual MSD among airport check-in workers involve:

- Lifting and carrying baggage;
- Fixed and constrained postures that are frequently awkward, uncomfortable, and maintained for too long a period. These conditions applied to check-in workers working predominantly in a sitting position as well

as in a standing position and, to a lesser degree, to those workers whose work station design encouraged a change of work position;

- Repetitive and sometimes forceful hand movements;
- An externally imposed pace of work; for example, a high volume of passengers, and at peak hours, a high pace of work.

Expediting passenger baggage involves entirely different demands. At airports with semi-mechanized baggage-handling systems, check-in workers have to manually lift and carry every piece of baggage checked in by passengers. The resulting workload and the risk for MSD, especially low back injury, can be likened to the MSD risks found in industrial workplaces where workers perform heavy manual lifting.

Findings by Reddell, Congleton, Huchingson, and Montgomery's investigation of airport baggage handlers indicated that weightlifting belts should not be recommended for check-in workers [58]. Further to their findings, it should be noted that weightlifting belts constitute personal protective equipment, which is the last line of protection against workplace exposures. Eliminating the exposure as far as possible is the first, and most protective, choice of measures. Check-in workers at semi-mechanized systems lift and carry every bag checked in yet received no training in health and safety, prevention, safe lifting technique, work station design options, nor are they hierarchically structured to be able to discuss options easily with management. It would be negligent to introduce personal protective equipment before first exploring more protective options.

Fully mechanized baggage-handling systems are meant to eliminate manual lifting and carrying of baggage, but there are problems associated with these systems as well. Frequent bending, twisting, squatting, and other awkward full-body motions while tagging baggage caused fatigue and created the risk for discomfort and subsequent injury.

At the airports in Québec and Switzerland, where a fully mechanized system was designed to eliminate manual lifting and work station design reduced full body motions, the risk for MSDs, including low back injury, still existed. There, postural discomfort, low-back pain, and musculoskeletal injury resulted largely from prolonged sitting or standing, and ill-adapted bodily motions (direct outcomes of the work station and work process), as well as from the manual lifting and carrying of passengers' baggage, which took place in varying degrees at all systems of mechanization.

The negative effects of these body-loading movements (including static posture imposed by work station design) and tasks were further exacerbated by the low level of control that check-in workers had over the speed, rhythm, and load demand of their jobs, general lack of training and skills development, and lack of involvement in the design of their work processes and work organization. The higher prevalence of upper and lower back pain at the Ontario airport provided evidence supporting the need for fully mechanizing baggage check-in systems. Eliminating the need for workers to lift and carry each checked bag is an important first step toward preventing work-related MSDs and injury in check-in workers. It is also significant that no check-in worker at any airport was provided with lifting aids or training in safe lifting techniques by management. Given the actual and potential loads on check-in workers' bodies, the significant rates of MSDs revealed were not surprising.

Fully mechanizing check-in systems, establishing a universal 20-kilogram-per-bag maximum limit, and providing worker training on safe lifting techniques do not necessarily guarantee the elimination of musculoskeletal injury. Such measures, however, constitute important pro-active steps. There is the greater need to examine the check-in job as an entire work system, rather than by isolated tasks, including examining the associated management practices. Biomechanical work factors, psycho-social factors, including the chronic distress, anxiety, and tension present due to the de-professionalization and de-skilling of the job, lack of value and consideration given to check-in workers' jobs evidenced by poor or complete lack of communication with management, the emotional labor associated with the job [59], violence on the job, and organizational factors appear to interact, contributing to the existence of or severity of MSDs and injury.

Factors that must be considered and addressed include quick turnaround policies, shift length, work load distribution throughout a shift, training and skills development, high work demand with a low level of worker decision latitude, strain from dealing with aggressive passengers and the ever-present threat of violence, including from abusive management. Input from check-in workers themselves may shed light on how best to diminish or eliminate the effects of these factors.

Should not employers of check-in workers, or third parties, such as manu-facturers of baggage check-in systems, fear expensive lawsuits from the onset of disabling MSDs, due to poorly designed work conditions? The lawsuit trend for work-related injuries and illnesses is ever increasing in a number of countries, most notably and expensively in the United States and the United Kingdom, where lawyers are the only ones consistently coming out on top. If workers do not register their work-related health problems, if communication remains poor between workers and management, then management and third parties may have no real need to fear lawsuits from workers, and the costs of worker health may continue to be externalized.

A systems approach is useful in research aimed at identifying the causes of work-related adverse health in workers. Of equal use-value and importance is the identification of suggestions for improvements or solutions to known or suspected problems. However, it is the joining of both strategies—identifying the root causes of problems and making studied suggestions for eliminating the problems—that is important for protecting workers' health and improving productivity and efficiency.

Applying a systems approach to prevent MSDs in check-in workers requires an understanding of key sources of stress, one of which is work-related violence, discussed in the next chapter. An additional important source of stress resulting in ill health is abusive managers and the particular forms of violence they inflict; issues which are addressed in chapter 5. With an understanding of the problems and their contexts in hand, one can then develop preventive measures. A series of detailed recommendations for improving check- in work through the prevention of MSDs and work-related violence is presented in chapter 6.

ENDNOTES

1. There are also some North American researchers who apply a systems approach in their work, examining cognitive and psycho-social factors together with work analysis. A number of those studies are referenced throughout this book.
2. B. Ramazzini, De Morbus Artificum, 1706, in *Europe Under Strain,* R. O'Neill (ed.), European Trade Union Technical Bureau for Health and Safety, Brussels, p. 31, 1999.
3. European Agency for Safety and Health at Work, *Monitoring the State of Occupational Safety and Health in the European Union—Pilot Study,* Office for Official Publications of the European Communities, Luxembourg, 2000.
4. R. Op De Beeck and V. Hermans, *Work-Related Low Back Disorders,* European Agency for Safety and Health at Work, Bilbao, 2000.
5. A. Karjalainen and S. Virtanen, Evaluation of the 1995 Pilot Data, in *European Statistics on Occupational Diseases* (EODS), 1999.
6. European Commission, *JANUS*:20-11, 1995.
7. National Academy of Sciences, *Musculoskeletal Disorders and the Workplace: Low Back and Upper Extremities,* Washington, D.C., p. 364, 2000.
8. U.S. Bureau of Labor Statistics, 1997.
9. T. F. Morse, C. Dillon, N. Warren, C. Levenstein, and A. Warren, The Economic and Social Consequences of Work-Related Musculoskeletal Disorders, The Connecticut Upper Extremity Surveillance Project (CUSP), *International Journal of Occupational and Environmental Health, 4,* pp. 209-216, 1998.
10. F. Tissot, K. Messing, and S. Stock, Standing, Sitting and Associated Working Conditions in the Quebec Population in 1998, *Ergonomics, 48*:3, pp. 249-269, 2005.
11. M. Kerr, J. W. Frank, H. Sh. Shannon, R. W. K. Norman, R. P. Wells, W. P. Neumann, C. Bombardier, and the Ontario Universities Back Pain Study Group, Biomechanical and Psychosocial Risk Factors for Low Back Pain at Work, *American Journal of Public Health, 91*:7, pp. 1069-1075, 2001.
12. S. Koda, S. Nakagiri, N. Yasuda, M. Toyota, and H. Ohara, Follow-up Study of Preventive Effects on Low Back Pain at Worksites by Providing a Participatory Occupational Safety and Health Programme, *Industrial Health, 35*:2, 1997.
13. National Institute for Occupational Safety and Health, *Cumulative Trauma Disorders in the Workplace,* NIOSH, U.S. Department of Health and Human Services, NIOSH Publication No. 95-119, 1995.
14. L. Rosenstock, *The Science of Occupational Musculoskeletal Disorders,* National Institute for Occupational Safety and Health, Report of written testimony submitted

to the sub-committee on work force protections, committee on education and the work force, U.S. House of Representatives, July 1997.

15. P. G. Dempsey, A. Burdorf, and B. S. Webster, The Influence of Personal Variables on Work-Related Low-Back Disorders and Implications for Future Research, *Journal of Occupational and Environmental Medicine, 39*:8, pp. 748-759, 1997.

16. W. Gereluk and L. Royer, *Sustainable Development of the Global Economy: A Trade Union Perspective,* SES Paper No. 19, International Labor Office, Geneva, 2002.

17. J. Rinehart, C. Huxley, and D. Robertson, *Just Another Car Factory? Lean Production and Its Discontents,* ILR Press, Ithaca, p. 71, 1997.

18. K. Hyytiainen and A. Uutela, Work Factors Related to Stress and Recurrent Low Back Pain Among Finnish Planners and Workers, *Work and Stress, 5*:3, pp. 197-204, 1991.

19. European Foundation for the Improvement of Living and Working Conditions, *Third European Survey of Working Conditions*, Luxembourg, 2000.

20. European Foundation for the Improvement of Living and Working Conditions, *Second European Survey of Working Conditions*, Luxembourg, 1996.

21. P. A. Landsbergis, J. Cahill, and P. Schnall, The Impact of Lean Production and Related New Systems of Work Organization on Worker Health, *Journal of Occupational Health Psychology, 4*:2, pp. 108-130, 1999; A. Statham and E. Bravo, The Introduction of New Technology: Health Implications for Workers, *Women and Health, 16*:2, pp. 105-129, 1990; J. V. Johnson, Occupational Stress, in *Preventing Occupational Disease and Injury* (2nd Edition), J. L. Weeks, B. S. Levy, and G. R. Wagner (eds.), The American Public Health Association, Washington, D.C., 2004; Hyytiainen and Uutela, 1991.

22. P. Leino, Symptoms of Stress Predict Musculoskeletal Disorders, *Journal of Epidemiology and Community Health, 43*, pp. 293-300, 1989.

23. S. Snook, The Practical Application of Ergonomics Principles, *Minesafe International*, 1993.

24. E. B. Faragher, M. Cass, and C. L. Cooper, The Relationship Between Job Satisfaction and Health: A Meta-Analysis, *Occupational and Environmental Medicine, 62*, pp. 105-112, 2005.

25. O. Heran-Le Roy and N. Sandret, Résultats de l'enquête Sumer 94: Les Contraintes Articulaires Pendant le Travail [Results of the Investigation Summer 94: Articular Constraints During Work], *Documents Pour le Médecin du Travail, 71*(3e trimestre), 1997.

26. L. Doyal, *What Makes Women Sick? Gender and the Political Economy of Health*, Macmillan Press, London, 1995.

27. S. Melamed, I. Ben-Avi, J. Luz, and M. S. Green, Objective and Subjective Work Monotony: Effects on Job Satisfaction, Psychological Distress, and Absenteeism in Blue-Collar Workers, *Journal of Applied Psychology, 80*:1, pp. 29-42, 1995.

28. J. M. Stellman, and S. Daum, *Work Is Dangerous to Your Health: A Handbook of Health Hazards in the Workplace and What You Can Do About Them*, Pantheon, New York, 1973.

29. J. M. Stellman, The Hidden Health Toll: The Cost of Work to the American Woman, *Civil Rights Digest, 10*, pp. 32-41, 1977.

30. T. L. Guidotti, C. Barbanel, and K. Auerbach, Occupational Medicine: The Case for Reform, *American Journal of Preventive Medicine, 30*:2, p. 186, 2006.
31. A. Aittomaki, E. Lahelma, P. Leino-Arjas, and P. Martikainen, Gender Differences in the Association of Age with Physical Workload and Functioning, *Occupational and Environmental Medicine, 65,* pp. 95-100, 2005.
32. M. Torgén and A. Kilbom, Physical Workload Between 1970 and 1993—Did It Change? *Scandinavian Journal of Work, Environment and Health, 26*:2, pp. 161-168, 2000.
33. K. Messing and D. Elabidi, Desegregation and Occupational Health: How Male and Female Hospital Attendants Collaborate on Work Tasks Requiring Physical Effort, in *Policy and Practice in Health and Safety,* IOSH Services Limited, pp. 83-103, 2003.
34. K. Messing (ed.), *Integrating Gender in Ergonomic Analysis: Strategies for Transforming Women's Work,* European Trade Union Technical Bureau for Health and Safety, Brussels, 1999.
35. Statistics Canada, 1995 Statistics Canada, *Work Injuries 1992-1994,* Ministry of Industry, Ottawa, Canada, pp. 72-208, 1995.
36. S. S. M. Yeung, I. T. S. Yu, and K. Y. L. Hui, World at Work: Aircraft Cabin Cleaning, *Occupational and Environmental Medicine, 62,* pp. 58-61, 2005.
37. T. Scherzer, R. Rugulies, and N. Krause, Work-Related Pain and Injury and Barriers to Workers' Compensation Among Los Vegas Hotel Room Cleaners, *American Journal of Public Health, 95*:3, pp. 483-488, 2005.
38. K. Messing and A. M. Seifert, *Francophone Ergonomic Analysis as a Method for Combating Gender Inequality,* presented at the Fourth International Conference on Work Environment and Cardiovascular Diseases, Newport Beach, California, March 10, 2005.
39. K. Messing, *One-Eyed Science: Occupational Health and Women Workers,* Temple University Press, Philadelphia, 1998.
40. L. Punnett, J. Gold, J. N. Katz, R. Gore, and D. H. Wegman, Ergonomic Stressors and Upper Extremity Musculoskeletal Disorders in Automobile Manufacturing: A One Year Follow Up Study, *Occupational and Environmental Medicine, 61,* pp. 668-674, 2004.
41. L. Alfredsson, C. Spetz, and T. Theorell, Type of Occupation and Near Future Hospitalization for Myocardial Infarction and Some Other Diagnoses, *International Journal of Epidemiology, 14*:3, pp. 378-388, 1985.
42. P. Smith and C. Mustard, Injury Statistics Don't Tell the Whole Story of Women's Work-Health Issues, *Occupational and Environmental Medicine, 61,* pp. 750-756, 2004.
43. For a discussion of the differential difficulties that women face in obtaining compensation for MSDs, see K. Lippel, Compensation for Musculo-Skeletal Disorders in Quebec: Systemic Discrimination Against Women Workers? in *International Journal of Health Services, 33*:2, pp. 253-282, 2003.
44. S. Haynes, A. Z. LaCroix, and T. Lippin, The Effect of High Job Demands and Low Control on the Health of Employed Women, in *Work, Stress and Health Care,* J. Quick, R. Bhagat, J. Dalton, and J. Quick (eds.), Praeger Press, New York, 1987.

45. A. LaCroix and S. Haynes, Gender Differences in the Health Effects of Work-Place Roles, in *Gender and Stress,* R. Barnett, L. Biener, and G. Baruch (eds.), The Free Press, New York, 1987.

46. J. Fleishman, The Health Hazards of Office Work, in *Double Exposure: Women's Health Hazards on the Job and at Home,* W. Chavkin (ed.), Monthly Review Press, New York, 1984.

47. J. Stellman and M. Henifin, *Office Work Can Be Dangerous to Your Health,* Pantheon, New York, 1989.

48. S. Cohen and N. Weinstein, Nonauditory Effects of Noise on Behavior and Health, *Journal of Social Issues, 37,* pp. 36-70, 1981.

49. R. Karasek, Job Demands, Job Decision Latitude, and Mental Strain: Implications for Job Redesign, *Administrative Science Quarterly, 24,* pp. 285-308, 1979.

50. R. Karasek, D. Baker, F. Marxer, A. Ahlbom, and T. Theorell, Job Decision Latitude, Job Demands and Cardiovascular Disease: A Prospective Study of Swedish Men, *American Journal of Public Health, 71*:7, pp. 694-705, 1981.

51. J. Siegrist, Adverse Health Effects of High-Effort/Low-Reward Conditions, *Journal of Occupational Health Psychology, 1,* pp. 27-41, 1996.

52. J. Prado-Lu, *Organizational Factors That Are Associated with Hypertension Among Women Workers in Industries That Have Accommodated Information Technology,* presented at the Fourth International Conference on Work Environment and Cardiovascular Diseases, Newport Beach, California, March 10, 2005.

53. K. Messing, M. Randoin, F. Tissot, G. Rail, and S. Fortin, La Souffrance Inutile: La Posture Debout Statique dans les Emplois de Service [Useless Suffering: Static Standing Posture in Service Jobs], in *Travail, Genre et Sociétés, 12,* pp. 77-104, 2004.

54. M. J. Smith and P. C. Sainfort, A Balance Theory of Job Design for Stress Reduction, *International Journal of Industrial Ergonomics, 4,* pp. 67-79, 1989.

55. T. R. Waters, V. Putz-Anderson, and A. Garg, *Applications Manual for the Revised NIOSH Lifting Equation,* Publication No. 94-110, National Institute for Occupational Safety and Health, Division of Biomedical and Behavioral Science, 1994.

56. R. Firth, *The Primitive Polynesian Economy,* Routledge, London, 1939.

57. Airport A in Ontario, Canada was the airport where workers stand the most and where most of the workers were women. The fact that among all three study sites more women reported suffering from foot pain is likely also a factor of there being only 1 male worker at Airport A, the "standing airport," compared with the other 2 airports.

58. C. R. Reddell, J. J. Congleton, R. D. Huchingson, and J. F. Montgomery, An Evaluation of a Weightlifting Belt and Back Injury Prevention Training Class for Airline Baggage Handlers, *Applied Ergonomics, 23*:5, pp. 319-329, 1992.

59. See chapter 4 for a discussion of violence and emotional labor in check-in work.

Afflictions of Service Work: Psychological Distress, Violence, and Poor Management

STRESS AND MUSCULOSKELETAL DISORDERS: THE MIND/BODY COCKTAIL THAT LEAVES YOU SHAKEN, NOT STIRRED

> Stress can wreak havoc with your metabolism, raise your blood pressure, burst your white blood cells, make you flatulent, ruin your sex life, and if that's not enough, possibly damage your brain [1].

Work-induced stress is a prime suspect in the etiology or aggravation of a variety of conditions, including MSDs and cardiovascular disease. Such conditions impose an equally wide variety of negative effects on private life, job satisfaction, and productivity. While researchers have debated the conceptualization of occupational stress [2], the idea that may have the greatest acceptance relates to the "person-environment" fit formulation put forward by McGrath in 1970 [3],

> . . . a (perceived) substantial imbalance between demand and response capability, under conditions where failure to meet demand has important (perceived) consequences [McGrath, 1970 in [2, p. 252].

The "person-environment" fit formulation refers to an excess of demands in relation to one's capability to meet those demands (e.g., overload), and underload, underutilization, and understimulation, which is the other side of the same coin [4, 5].

In an effort to understand the etiologic processes by which conditions of work lead to ill health, Kasl and Cobb (1983) enumerated four useful characteristics of such processes:

1. The stressful work condition tends to be chronic rather than self-limiting.
2. Habituation or adaptation to the chronic situation is difficult, and some form of vigilance or arousal must be maintained.
3. Failure to meet the demands of the work setting can have serious consequences.
4. There is a "spillover" of the effects of work stress onto other areas of functioning—such as family and leisure—so that the impact of the demanding job situation becomes cumulative and health-threatening (rather than being daily defused or erased with consequently little long-term impact on health) [2, pp. 254-255].

A body of evidence today demonstrates a strong association between job strain and cardiovascular disease. Twenty years ago it was thought that personal characteristics strongly influenced disease outcomes from occupational strain, but today research shows that both personal and job characteristics are important in strain-related outcomes, while social support plays a critical role in buffering or protecting against the effects of stress and job strain [6]. The demand-control model [7, 8] predicted the importance of social support as a key resource in the work environment for reducing stress-related disorders. Strong control over one's work with a high level of social support has been associated with higher work satisfaction and lower instance of psychiatric distress [9]. Conversely, high psychological job demands, low work control, and low co-worker social support combined have been associated multiplicatively with cardiovascular disease prevalence [10].

If the impacts of job dissatisfaction on mental health and physical well-being are powerful, the impacts of job satisfaction are also strong. Clearly job satisfaction is an important objective for management practices to focus on improving. Job satisfaction is closely linked to work conditions. Characterized by Locke [11], desirable work conditions are where

> [t]he work represents mental challenge (with which the worker can cope successfully) and leads to involvement and personal interest; the work is not physically too tiring; rewards for performance are just, informative, and in line with aspirations; working conditions are compatible with physical needs and facilitate work goals; the work leads to high self-esteem; and management and co-workers are supportive [11, cited in 2, p. 257].

One cannot necessarily draw a direct cause-and-effect linear relationship between frustration, dissatisfaction, and illness. However, there are indications that the present trends dominating work in today's globalized society increasingly have negative effects on the psycho-social and physical health of working people [12]. Work-related stress is a major and growing problem worldwide. Does anyone still think that the mind and body are not closely connected, that the psyche does not affect physical health? Looking at the burden of the entire work system on workers' health and well-being—including the impacts of management practices and the effects of emotional demands—is, understandably, complicated research, and under the publish-or-perish pressure in academe, studies must be manageable in scope. It is easier to measure single factors or to analyze individual movements than to try to assess a multitude of factors and the political context that may contribute to work-related stress as well as how they might in turn contribute to physical ailments.

Nationally representative surveys of U.S., Canadian, and EU populations reveal 20-30% or more of the work force are often or always stressed in their work [13-15]. In the United Kingdom, an estimated 6.5 million work days were lost due to work-related stress in 1995, while the number increased in 2001 to an estimated 13.4 million work days lost due to work-related stress, depression, or anxiety. In 2001, individual workers lost on average 29 days due to occupational stress compared with an average of sixteen days lost per person in 1995 [16]. Estimates for the United States showed that more than 28% of hypertension among working men in New York may be attributed to stressful work conditions, characterized by job strain (the combination of high demands with low control) [17].

There seems to be general agreement that work-related stress can be understood to be a process, whereby the effects are cumulative over time with continued exposure. To date, the most consistent findings are that workers with low work control are at significantly greater risk of premature death, and that work stress has been linked to increased risk of MSDs, which are greater among workers with high physical demands [18].

Work with the public is associated with a particular form of *emotional* strain, characteristic of service-sector jobs usually performed by women. Emotional strain in service-sector work, because of what some researchers have referred to as "emotional labor" [19], is a result of the pressure to respond to the needs of clients, customers, or passengers, the need to control or change one's emotions to match those required by the company or organization. (Emotional strain is distinguished from job strain, the latter referring to the combination of high demands with low decision latitude.) Jobs that involve selling one's emotional labor require workers to pretend to have positive feelings that they do not really feel and to deny their negative responses in order to make clients, customers, or passengers feel cared for. Workers may feel what they have described as a "loss of self" when feelings and emotions become dulled from defending themselves against difficult situations with passengers, clients and customers [19]. Women with the least control over their work lives, (often those in low status jobs), suffer the most negative effects of stress-related disorders, including those from emotional labor [20]. The same association has been made for male-dominated jobs, where workers have little or no control over their work lives.

The emotional strain created by selling one's emotional labor is particularly relevant to the work of airport check-in workers where the center of control may pass from the worker to the passenger or customer. Check-in workers have to be friendly and helpful to every passenger, treating each one as special, regardless of how many customers they already have checked in on a given shift in a given day. Check-in workers have to appear cheerful at all times regardless of the fact that they may be looking at a long crowded line of impatient people waiting to be checked in with speed, efficiency, and a friendly face. The special needs presented by any number of passengers can be particularly taxing to the nerves of check-in workers and may require a high degree of emotional control. A low level of worker control over his or her emotional labor combined with a high-demand job creates additional negative stress. "Women are hired in the labor market as women in their nurturant, supportive, and maintenance (or domestic-type) services" [21]. Although written in 1980, Natalie Sokoloff's analysis is no less true now and is particularly apparent in women's service-sector jobs and white-collar jobs where they most often are subordinates of men.

Holte, Vasseljen, and Westgaard used a combination of instruments to identify episodes during a work day that resulted in physical responses to elevated tension [22]. Upper body muscular tension and increased heart rate were responses to a variety of situational demands; a common factor was contact with other people who caused negative emotions (for example, aggressive passengers, in the case of airport check-in workers). A significant relationship was found between muscle tension and emotional state, indicating that negative emotions contributed to the development of tension, where tension, more than a stressful environment, was associated with negative emotional responses [22, p. 131]. This study demonstrated a physiological association with the subjective perception of tension in workers, pointing to the importance of possible underlying pain-inducing mechanisms that would be missed by traditional standards used for recording and evaluating muscle responses to work exposures causing tension.

As with other characteristics of women's work, the emotional strain involved in work with the public has been little researched, despite stress, exhaustion, and "burn-out" being common among service-sector workers. A study of psycho-social risk factors to health in women telephone operators and sales agents in northern Mexico found that workers must endure high levels of emotional labor, stress, and violence from customers. Emotional labor is in the form of emotional demand as a requirement of the job, introduced by the company during job training. A high degree of chronic stress is created by verbal aggression from telephone customers, with more than 80% of the studied population experiencing it. From 30-40% of daily calls constituted sexual aggression to workers; insults were experienced as well [23].

Emotional strain associated with service-sector work performed by women [19], viewed through a lens of the high demand/low decision latitude job category, fills out a picture of check-in work as a potentially highly stressful job.

[8]. These objective psycho-social factors, associated at once with the very nature of the job as it was conceived and with the subjective experiences with passengers, are thus supposed significant causal factors in MSDs. Such factors should be considered, regardless of whether a check-in system is semi-mechanized or fully mechanized.

Women's suffering is less visible than men's, hidden behind what appear to be easy and clean work conditions. A wide range of physical and psychological symptoms, including muscular tension, often accompany emotional distress and anxiety, while the etiology of symptoms frequently is unclear. One can imagine that workers experiencing high levels of work-related stress may be more likely to report MSDs when asked [24], although it is an area that has not been investigated sufficiently among women workers in service-sector jobs.

Compounding the negative stress associated with the emotional labor of service-sector work and the stressful nature of jobs characterized as high-demand with low-control, violence in the workplace is an additional and growing factor contributing to work-related stress. One need not stretch the imagination to understand that negative stress-inducing psycho-social factors work in combination with physical and organizational factors causing MSDs in check-in workers, as may be the case in any group of service-sector workers.

Work activities characterized as repetitive, oversimplified, low in status, low in sense of control and in sense of participation, tend to be associated with the worst physical and job satisfaction dimensions. Strong associations have been found between psycho-social aspects of the work environment, in particular imbalance between effort and reward at work, and depressive symptoms [25, 26]. A particular association has been established between work conditions, particularly stress-induced adverse health outcomes, and sickness absence among women workers, and it has been linked directly to monotonous work. Characteristics signaling work underload, such as employment beneath one's capacity, too narrow job content, lack of task variation, lack of demand on worker creativity or problem-solving capacity, and insufficient social interaction are considered as stressful as hectic repetitive work. Work underload and repetitive work create negative stress, have been linked to MSDs through chronic strain, distress and muscular tension, and are both preventable. Monotonous repetitive work is typically accompanied by the non-involvement of workers in any level of organizational decision-making, including decisions directly affecting workers' jobs.

The examination of studies of computer clerical workers, supermarket check-out workers, and airport baggage handlers revealed that the mainstream research approach is more linear, or single factor, than multi-factorial, in nature. The growing challenge of stress related to work necessitates more than ever the application of a "holistic" approach in occupational health research and practice that focuses on the overall work environment rather than narrow individual risk factors in work.

Among all three of the surrogate groups, studies that led to workplace improvements and an organizational change process needed to sustain improvements had a systems approach in common, taking into account the totality of the work system, and applied a multi-variable and multidisciplinary approach to job examination. Intervention-oriented studies on computer clerical workers and supermarket check-out workers argued convincingly for the examination of psycho-social factors, physical factors, and organizational factors in combination when looking at causes of MSDs. "Organizational factors" present perhaps the greatest challenge to address at the workplace level because indisputably the outcome of any such examination must indicate the need for change in management practices if the employer is truly committed to the promotion of worker health and well-being, and if the employer is committed to the use of workers' full potential.

Management practices are largely responsible for the high prevalence of work-related violence, whether in the form of passenger violence against check-in workers, or exhibited as managerial abuse toward workers, or generated by any other person or group. Both employers and workers pay the price of work-related violence when systems of prevention and protection do not exist within a company or organization. Such systems embody raising awareness at all levels, including about the characteristics of destructive personality types.

VIOLENCE AT WORK

"I hope not to be doing this work for too much longer," a worker at Airport B said. "I preferred being a flight attendant—there were less aggressive passengers." Other workers expressed similar feelings, citing supervisors and upper level managers as the source of violence against check-in workers. Violence against workers comes in numerous forms and has consequences for both workers and employers. The effects of physical violence against workers are wide ranging, with the most serious outcome being suicide, while verbal abuse and other less extreme forms of violence also have serious consequences. The evidence has mounted demonstrating that various types of violence represent problems of a disturbing magnitude that affect working people around the world. A considerable number of people are exposed to physical assault on the job. Violence against women in particular has started to gain the attention needed to capture its extent and magnitude. In the workplace (and outside it), violence against women is not only a manifestation of gender inequality, but is a means of maintaining an unequal balance of power that is used to subordinate women [27]. In the workplace, any sort of harassment (harassment constitutes abuse)—psychological, physical aggression, sexual harassment—constitutes workplace violence. Whether the abuse is from those in positions of authority or from aggressive passengers in the case of airport check-in workers, violence at work should be seen as a violation of one's human rights. Exposure to workplace violence often has dramatic impacts on those exposed directly and indirectly; the impacts are felt both by individuals and enterprises. The consequences of workplace violence often result in time off from work and reduced productivity when affected individuals are at work [28].

Evidence further suggests that workplace violence is increasing, undoubtedly linked to recent socio-economic trends and societal factors that include increasing competitiveness, work intensification, increased emotional demands, increased control over workers, reduced worker autonomy, increased work time and "time squeeze" in general, and downsizing. These factors all create pressure, stress, strain, and insecurity for working people everywhere. With the cultural and structural hegemony of work in modern times consisting of work intensification, an increased speed of work, the drive for ever greater productivity, fiercer and fiercer competition, producing more with fewer people, and the reduction and elimination of benefits for many working people, an increase in workplace violence should come as no surprise [29].

In the meantime, there is a slow-growing proliferation of workplace-based training programs to address violence at work. Many programs may appear palatable by the way they are marketed. In reality, many such programs, no matter how they are packaged and no matter what seemingly credible group, organization, or company is behind them, are focused (either outright or thinly veiled) on the individual and not aimed at organizational change. Workplace policies that address drugs, alcohol, stress, and violence are merely band-aids if they do not tackle the underlying management and work practices that are at the root of the problems in the first place. Such packaged programs and policies *cannot solve* the tackled issues without a *structural response*. A meta-analysis of 485 studies linking job satisfaction with physical and mental well-being concluded that organizations should include the development of stress management policies to *identify and eradicate work practices that cause most job dissatisfaction* [30].

Workplace policies are insufficient at best if they do not address the destructive management practices that lead to job dissatisfaction and often to the use and abuse of coping mechanisms (particularly drugs and alcohol), if they do not address the structural hegemony, if they do not address the lack of worker involvement in workplace decision-making, the work demand/control dichotomy, the manager/subordinate hierarchical dichotomy, and declining social support. Indeed, social support appears to be incompatible with fierce competition, downsizing, labor flexibility, and increased insecurity. For example, both depression and alcohol dependence have been correlated with high work demands and low control [18]. Even worse, workplace violence and stress programs and policies run the risk of placating workers when they are presented so that workers will think management is doing something about the problem by introducing training and organizational policies.

The body of evidence shows that successful primary prevention of work-induced stress has focused on job control and social support available in the workplace. Secondary prevention has focused on coping abilities [18]. Secondary prevention interventions are effective, but only among those with a high degree of workplace control. Training and organizational policies are necessary and useful, but only if they address the root causes of the problems and do not end up "victim-blaming" employees who do their best with increasingly unsustainable and intolerable work conditions. Such policies and efforts must focus first and foremost on primary prevention, which means removing the sources of stress, rather than trying to get working people to adapt to the stressors.

The current hegemonic trend, combined with a focus on the individual and away from the collective, is presumably responsible in large part for aggression at work seeming to be the new model replacing respect. If respect at work was status quo in decades gone by, what often appears to be a total absence of respect seems to dominate the world of work today in Western industrialized countries. It is within this perspective that working people need to consider the increase in workplace violence. Accepting to be treated without respect at work must be seen, and dealt with, as entirely unacceptable. The necessary response to the cultural and structural hegemony is a call for collective action, for strong collective voice, for a widespread "raising of the bar" [31] and refusal to accept the unacceptable.

In the United States, homicide is the second highest overall cause of workplace deaths and is the leading cause of workplace death for female workers [32]. Workplace violence accounts for 15% of the more than 6.5 million violent acts experienced by residents of the United States aged 12 or older [33]. Each year in the United States, approximately two million individuals become victims of workplace crimes, resulting in considerable economic losses. Findings from a 1992-1993 U.S. Northwestern National Life Insurance Company study showed that 2.2 million full-time workers had been physically attacked on the job, 6.3 million had been threatened with violence, and 16.1 million had experienced harassment on the job. And those were only the reported cases. Management reports indicated that significant numbers of workers were harassed by co-workers, bosses, customers or clients, strangers, and former employees [34]. These problems are not found only in the United States—the Canadian Union of Public Employees has found verbal aggression to be the leading form of violence against employees [35].

The costs of workplace violence are high to workers, their families, employers, and society at large. Some researchers have estimated violence at work and its associated effects to account for 1-3.5% of GDP [36]. The total costs of work-related violence to organizations are difficult to measure, since many of the effects are hidden or not obviously associated with incidents of violence, particularly those related to productivity, morale, and absenteeism. In many countries, violence is increasing most rapidly in the service sector, with growing demands and pressures from clients and customers affecting a significant proportion of the working population. This problem does not spare the service job of airport check-in workers.

It has been estimated that workplace crime victimization costs employers about 1,751,100 lost workdays per year, averaging 3.5 days per crime, resulting in $55 million per year in lost wages, not including the cost of sick days and annual leave [33, p. 8]. Incidents of workplace violence have been estimated to cost companies more than $4 billion in lost work and legal expenses; a single episode may cost employers an estimated $250,000 in lost work time and legal expenses [34, p. 68].

Despite these alarming data, few enterprises or organizations have any kind of violence prevention program or a plan to deal with the aftermath of a violent incident. Findings from the U.S.-based Society of Human Resource Management indicate that less than 30% of companies have a formal pro-active plan to prevent violence or deal with its aftermath and even fewer have plans to introduce such a strategy. This lack of preparation appears to be the result of a lack of knowledge of what to do, cost factors, and a belief that violence will not happen. Yet the incidents do take place, and workers are on the front line to bear the brunt of the incidents.

Airport check-in workers are on the front line of the aggressions and attacks of many traveling passengers. Yet the employers of airport check-in workers do not have violence prevention programs in place, nor do they have any well-structured plans for dealing with the aftermath of violent incidents. It is a problem. But it is a problem that can be solved.

Workplace violence is also a growing problem in Europe. Within the EU, interventions aiming to address workplace violence issues have been introduced, but there are indications that the interventions are problematic [37]. Nonetheless, the existence of interventions shows that some organizations are working toward preventing and alleviating violence at work, although progress often is slow. Interventions in some organizations have led to major cost savings, however the bottom line should not be the primary driver. The evidence should stimulate companies and organizations to implement intervention measures and to monitor their effects [28, 30-40].

While workplace violence mostly has been identified as a problem of indust-rialized nations, evidence suggests that it is a problem of a global nature, affecting workers in developing as well as industrialized regions. There is no reason to think that psychological or physical violence on the job is limited to some cultures and not others. Although few data are available for developing countries, there is evidence of increased awareness of the problems associated with violence, for example, in India [41] and Mexico [42]. ILO People's Security Surveys show that workers in numerous developing and transitional economy countries report being exposed to harassment or violence on the job; women are the most affected [43, 44].

The prevalence rates of physical assault and sexual violence in the workplace are summarized in Table 1, by regions around the world.

Individual costs of workplace violence are difficult to quantify, including the costs of the personal suffering and pain that result. Suffering and humiliation are not self-contained events and usually lead to lack of motivation, loss of confidence, and reduced self-esteem, irritability, depression, anger, and anxiety [46]. These indicators typify stress, which can turn into a very costly matter. If the causes of violence are not eliminated or the effects not contained by adequate intervention, the symptoms are likely to develop into physical illness, psychological disorders, tobacco and alcohol abuse, or other negative coping strategies. Additionally, they can culminate in occupational accidents, invalidity, and even suicide.

If one takes a broad view of the cost of workplace violence, economic as well as non-economic costs should be included. Individual costs include loss of income from sickness absence, a cycle of eventual medication and medical consultations, and exit from work or retirement. Additional costs include impacts on private and social life, including intangible effects such as the cost of suffering or personal loss. Organizational costs primarily relate to sickness absenteeism, reduced productivity, replacement costs, and additional retirement costs. Further costs may be due to damage in production or equipment and a potential public loss of goodwill also may be incurred. Costs to society include medical costs and possible hospitalization, benefits and welfare costs connected with premature retirement, and the loss of productive workers.

Researchers have shown that the ramifications of workplace violence or work-related accidents extend beyond the individuals personally assaulted, attacked, or

Table 1. Prevalence Rates of Physical Assault and Sexual Violence at the Workplace, by Type of Incident, Sex, and Region, 1996 (percent of labor force)

	Assault		Sexual incidents
	Male	Female	Female
Western Europe	2.7	3.0	5.4
North America	2.5	4.6	7.5
Eastern Europe: Countries in transition	2.0	1.4	3.0
Asia	0.4	1.0	1.3
Africa	2.3	1.9	3.7
Latin America	1.9	3.6	5.2

Source: International Crime Victim Survey (ICVS) [45].

injured [47]. Those witnessing violence may live in fear of future violent incidents and experience negative effects similar to those of the workers who were the direct victims of the assault, attack, or aggression. A violent incident may generate fear in the group about the overall safety of the occupation. The fear, distress, anxiety, pain, and suffering related to violent incidents or accidents are not effects experienced only by women—both men and women experience the psychological and physical ramifications, albeit in a different manner [48]. How groups in a workplace deal with the aftermath of such events also differs.

Unspoken rules may exist among some work groups, for instance, not complaining about aches, pains, fears, anxieties, general physical health problems, or psychological health problems, as a means of maintaining group cohesiveness. Airport check-in workers appear to be a work group with common characteristics of identification, where pain and suffering are the norm, and where violence from aggressive passengers is a recurring threat, one known to check-in workers (see Figure 2). Retail and service industries are the workplaces that have the highest risk of violence due to cash transactions and accessibility by the public. Despite the high costs to organizations and society of workplace violence, the knowledge of violence, particularly about non-fatal acts of violence, is limited and the true overall costs have not been measured [49].

For airport check-in workers, direct accessibility by the public is a significant risk factor for violence on the job. The risk of what has come to be known as "ground rage" is high, and is an area of major concern of the ITF, highlighting once more the need for airport check-in workers to be considered as safety professionals on the ground.

VIOLENCE AT THE CHECK-IN:
GROUND AND AIR RAGE

If airports, handling agents and airlines don't treat their staff as safety professionals, it can hardly come as a surprise that passengers don't do so either. The management of unruly or disruptive behavior can best be achieved where passengers recognize the safety role of staff, are willing to accept their authority, and will comply with their instructions. Unfortunately, the marketing of aviation very often explicitly undermines this staff role: crew and passenger handling staff are all too often portrayed as compliant service providers, willing and able to meet the individual requirements of passengers. Service with a smile, delivered by young, attractive, and usually female staff is the standard approach when promoting airline brands. Such images promote in the mind of passengers the notion that crew and ground staffs exist to meet passenger demands alone, rather than to enforce and deliver passenger safety [50].

Violence against check-in workers from aggressive, disruptive, or unruly passengers is a real and growing phenomenon [51]. Compounding the already stressed and nervous state of many passengers on arrival at an airport, any number of whom may fear flying, airlines today are competing increasingly for loyal customers, using promises of seat location, meal choice, mileage benefits, on-time departures, on-time connections, baggage security, and offers of low-priced tickets. These factors contribute to customer demands and passenger feelings of entitlement that are expressed at the check-in counter, the first point of encounter between traveling passengers and airport staff.

The problem of aggressive or disruptive passengers traveling by air has markedly increased in the past several years, including threats, verbal abuse and physical abuse from them [52]. Frustration at the lack of progress in dealing with this growing problem prompted the ITF to launch in 2000 and 2001 a worldwide campaign against air rage to protect cabin crew. The ITF estimated that there was a fivefold increase in incidents from 1994 (1,132 incidents) to 1997 (5,416 incidents). The U.S. Aviation Safety Reporting System indicated that reports of aggressive passengers increased by some 800% between 1997 (66 incidents) and 1999 (534 incidents). These rates of increase may reflect increased reporting, but most likely they also reflect underestimates, because reporting typically included only actual violence but seldom addressed threatening behavior. As a result of lobbying by U.S. trade unions, in March 2000 the U.S. government increased the maximum fine for disruptive passenger behavior offenses from $1,100 to $25,000 [51].

> A few years ago, a man reached over the check-in counter and grabbed a woman by her hair when she was working at the check-in in Amsterdam Airport. He yanked her over the counter by her hair and hit her in the face. There are more and more of these kinds of incidents. (This incident was experienced by a worker of one of the ITF's European affiliates and was described to me by the Head of the ITF's Civil Aviation Section, London.)

Given the lasting impact of witnessing such violence on the job, it is not surprising that check-in workers did not particularly wish to describe in detail violent acts that they had experienced or witnessed on the job. Crull showed that anger, fear, depression, embarrassment, guilt, and sleeplessness were frequently reported responses to aggression (sexual harassment in particular) against women, and that fear and anger often interfered with the full range of personal relations [53]. Excessive tension and symptoms of psychological distress were found in a majority of women who had faced acts of aggression.

There is a need for multi-level action to combat the problem of workplace violence against workers, including check-in workers and other ground staff. Action is needed at individual, group, and organization-wide levels [47, 54, 55], including prevention, intervention, and rehabilitation. Organizational support helps to diminish the negative effects of violence against workers when incidents do occur, by buffering the negative effects in a way that support from family and friends does not provide [47].

An employer attitude of zero-tolerance for work-related violence, including all forms of psychological violence—abusive or disrespectful treatment from managers or co-workers, harassment, mobbing, and bullying—is a necessary organizational focus, which should be communicated to and by Human Resource Departments. Of equal importance is for respectful treatment to be demonstrated by top management and communicated to all employees. Policies and procedures, security, crisis management teams, and employee assistance programs, as well as educating workers and managers about work-related violence, are necessary components of an organizational approach aimed at zero-level work-related violence [56].

Data on workplace violence are becoming more common. Employers intending to develop and maintain productive and healthful work environments must protect themselves, their workplaces, and employees to minimize risks from potentially hostile sources. A positive sign is that more human resource managers and their employers in the public sector are addressing the rapidly changing employment arena by establishing workplace violence prevention plans and devising related training for all employees. Airports around the world would do well to do the same. Today, and even more so since the events of September 11, 2001 in the United States, it is crucial that workers responsible for public safety be specially trained, particularly in airports and other public transport facilities.

Airport check-in workers, the first to make contact with traveling passengers, have a wide range of skills that can be used in the role of safety professional on the ground. They are a logical first line of defense for public safety in airports. It is important that check-in workers have the confidence, commitment, and authority to deal with aggressive and disruptive passengers [50]. Given the necessary training and management support empowering them to take on this critical role, check-in workers are well placed to identify potentially aggressive or threatening passengers and to catalyze a chain of actions aimed at preventing disasters or aggressive behavior toward check-in and other air transport workers and passengers. Recognition of the positive and important role that check-in workers can play as safety professionals on the ground would be a natural means of broadening their skill base, providing well-deserved recognition to the professional job performed, and it would be a way to offset the cyclical and repetitive nature of some of their job tasks.

Check-in workers are well placed to assess the weight of bags, to ensure that carry-on baggage is correct in weight, size, and number; identify passengers who are under the influence of alcohol at time of check-in (before they become a potential problem on board an aircraft); and to identify passengers with mood problems (also before they become a potential problem in flight), all of which are factors identified as potential causes of air rage. Such skills are key means of preventing passenger rage on the ground or in the air, but are not currently being exploited by the airlines or airport management authorities.

Systems for recognition and action aimed at the prevention of workplace violence do not yet appear to be commonplace for the protection of airport check-in workers. The recent focus on aggressive passengers traveling by air has been aimed at the protection of cabin crew and passengers and has not involved ground staff. Security workers are an additional group receiving attention as a direct outcome of the September 11, 2001, attack in the United States. Violence against workers, including the threat of violence, produces stress. A wide range of both somatic and psychological effects is demonstrated to result from work-induced stress, including from the threat of violence or perceived threats [57]. Muscular tension is one of the symptoms often accompanying emotional distress and fear. Combined with the daily, even hourly, pressure to be ever cheerful and helpful to passengers, it is reasonable to assume that the muscular tension associated with work-related violence or the perceived threat of violence, compounded by the emotional labor associated with their work, are factors contributing to MSDs in check-in workers beyond the more obvious bio-mechanical factors.

Some airport check-in workers told specifically about their experiences with violence at work. Violence at the check-in can take the form of verbal abuse, physical threats while working, and outright physical assault or attack. The study team asked workers to what degree they considered that violence at their job represented a risk. The findings were revealing, both in terms of types of violence experienced by check-in workers and the level of perceived risk of facing aggressive or violent passengers (Figure 1). Interviews and focus group discussions confirmed verbal abuse and violence in general as primarily

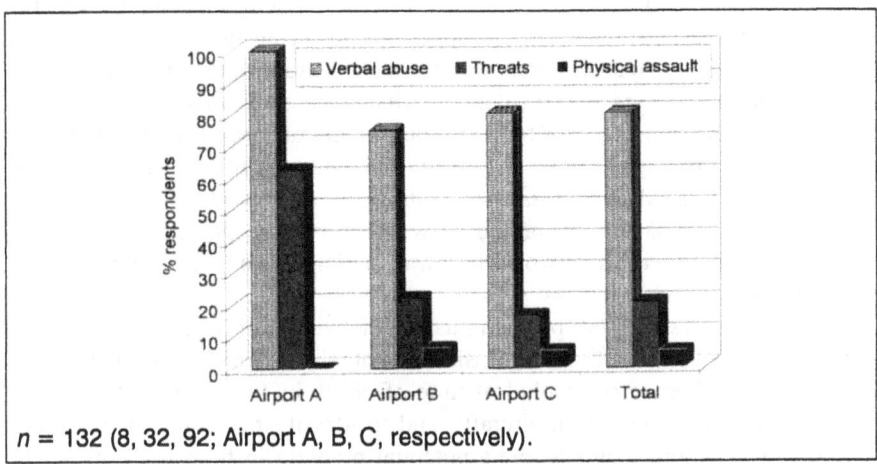

n = 132 (8, 32, 92; Airport A, B, C, respectively).

Figure 1. Experience with violence (% respondents having
experienced specified type).

a passenger-related problem, while problems of abuse and violence from supervisors and co-workers arise as well. Focus group discussions with ground staff around the world revealed instances of abusive treatment from management and negative impacts on workers from destructive management. In some cases, women workers said they were treated with a lack of respect by management.

> We were told even to hike up our skirts almost to the thigh, in order to sit on stools they bought and made us sit on, while being obliged to wear a skirt or dress as part of the uniform. We were not allowed to wear trousers—only men were. It was miserable, uncomfortable on those stools with no backs, and humiliating. And the men have fewer injuries than the women do, because we have to wear pumps while climbing over the belts, climbing on the baggage carousel, lifting bags, while the men can wear sturdy shoes. They are better protected when you trip or from things hitting your feet. (Ground staff worker)

> In the international flight area, passengers have easier access to us, so there is more danger with aggressive passengers. (Worker, Airport B)

> When I have to deal with abusive passengers, I call for a supervisor, but if I have to deal with an abusive co-worker, I probably would not report it. (Worker, Airport B)

Workers at all three airports reported that they had been subjected to at least one form of abuse from passengers. More than 80% had been subjected to verbal abuse from passengers, more than 20% had received threats from passengers, and more than one out of every twenty check-in workers had been physically assaulted on the job. These figures are for incidents experienced during their employment as a check-in worker. A number of check-in workers mentioned the difficulty of dealing with aggressive passengers and the associated stress, and highlighted that they did not receive support from management.

> It is hard to handle the difficult passengers, and you don't really get help. (Worker, Airport A)

> Abusive passengers are hard to deal with. I have approached my supervisor about this, but didn't get any support. (Worker, Airport B)

> If you witness violence, or if you're the victim of violence, the supervisor on the floor will be called. They call in security when the situation gets worse. Dealing with angry or disgruntled passengers, like when a flight is canceled or overbooked, is really stressful. We go out for a smoke to deal with the stress. (Worker, Airport B)

In addition to the act of verbal or physical violence, the perception of the potential to be a victim of violence or the perceived risk of exposure to work-related violence is revealing (Figure 2). More than half of the workers considered violence a low risk at their jobs, more than one-third perceived that their jobs presented a medium level of risk, and more than 5% saw their jobs as placing them at high risk of work-related violence. Taken together, nearly half of the workers (46.2%) perceived a substantial risk of violence in their work. These findings indicated the perceived risk of violence as significant in the job of airport check-in work [28].

The findings are alarming and should be considered in the context of workers performing a job every day with the underlying distress and anxiety of knowing that past experiences with violent passengers can be repeated at any time with new passengers. If not addressed, this chronic stressor has the potential to engender a feeling of victimization among workers whose job status and training do not empower them with tools to deal adequately with situations of violence or potential violence. The only choice is for workers to remain in a constant state of vigilance, one in which the body is always prepared to meet some physical or emotional demand.

The finding that more than 80% of workers (male and female) had personally experienced verbal violence and that more than 5% had been physically assaulted by passengers makes it likely that many more workers have witnessed violence against check-in co-workers, even if they themselves were not the target of the

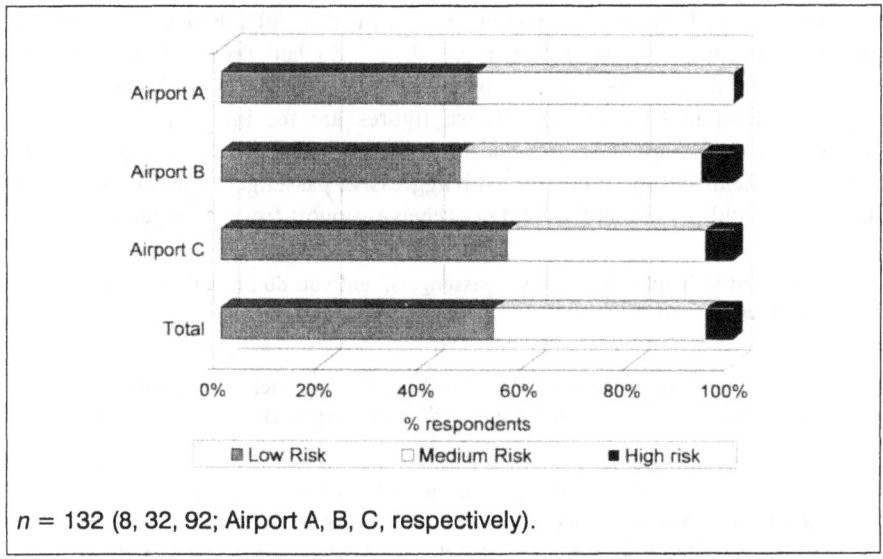

n = 132 (8, 32, 92; Airport A, B, C, respectively).

Figure 2. Perceived level of risk related to violence.

aggression. The ITF, check-in workers, and management at the three airports all reported that aggressive passenger behavior is increasing, both in the air and on the ground.

The underlying strain associated with facing aggressive passengers without tools or formal support systems for protective action is an important and constant stressor for check-in workers. Returning to the job after being subjected to any type of violent incident without any reactive or pro-active systems to address such incidents (even mild incidents), would, presumably, create anxiety or provoke a low-level constant fear. One of the responses associated with fear is muscle tension. Living with a level of fear, or apprehension on the job every day, is therefore a likely significant factor contributing to MSDs, compounding the biomechanical stressors found on the job.

Taking a pro-active approach to the problem, in 2000, management at the airport in Switzerland (including the two separate employers of check-in workers there) initiated a 2½-day training seminar about dealing with aggressive passengers, which was obligatory for all check-in workers, supervisors, and managers. A worker there commented, "Aggression training is new this year. It's a response to this problem increasing. Ten people attend a two and a half day course at a time." The seminar provided workers, supervisors, and managers with tools for dealing with aggressive passengers and gave the power and authorization to check-in workers to say "no" to aggressive passengers in certain situations, even though doing so had been against airline policy. Implementation of this kind of seminar, with mandatory participation by both workers and managers, was one step in a pro-active management practice and is an important means of helping to pre-empt incidents of aggressive behavior in the air, where the consequences can be more far-reaching. However, a one-time seminar does not constitute an organizational policy against violence and alone is insufficient, although it is a good first step. Unfortunately, in this case, it was the only step.

The root causes of passenger violence against check-in workers are likely multiple in nature but have not been studied. The subject would be an important area of investigation to undertake, as indicated by global affiliates of the ITF and by airport management. Some key factors contributing to passenger violence against check-in workers, as suggested by unions and management, include alcohol; drugs; anxiety associated with air travel; time pressure; delayed, canceled, or missed flights; fear of flying; sense of entitlement due to airline advertising and competition policies; increased numbers and types of people traveling; and experience on board aircrafts in confined spaces for long periods of time. There may be other less obvious factors as well.

Management at the Swiss airport suggested that the overriding factor causing passenger aggression might be the competitive policy of airlines, aimed at obtaining customer loyalty by making various promises through advertising and publicity. Passengers, in consequence, feel entitled to demand what was "promised" to them, making such demands first and foremost to airport check-in workers.

It is reasonable to assume that airline management policies have direct consequences that may increase passenger violence against workers. The stress and strain created for workers through these policies surely contribute to negative health outcomes, even beyond MSDs. The outcomes do not appear to result from one cause alone. The lack of protective and pro-active management policies is one factor in the equation; airline policies that contribute to passengers' demands of check-in workers are additional factors. There does not appear to be much end-utility in addressing any single factor in the absence of the others.

Figures 3 and 4 present check-in workers' experience with violence, by sex. The small number of male check-in workers meant that firm conclusions about men or women being more exposed to one type of violence or another cannot be drawn. It is clear from the findings, however, that men as well as women reported verbal abuse and threats. No men in the study said they had experienced physical assault by passengers—only women had been the victims of passenger-generated physical assault. Male check-in workers from all three airports saw check-in work as exposing them to a degree of risk of violence. Some male workers considered check-in work low risk, while others believed that it exposed them to a significant risk of violence. The fact that violence was experienced by check-in workers at three different airports in two countries known for peaceful civilian life (Canada and Switzerland are countries known for their low levels of violence)

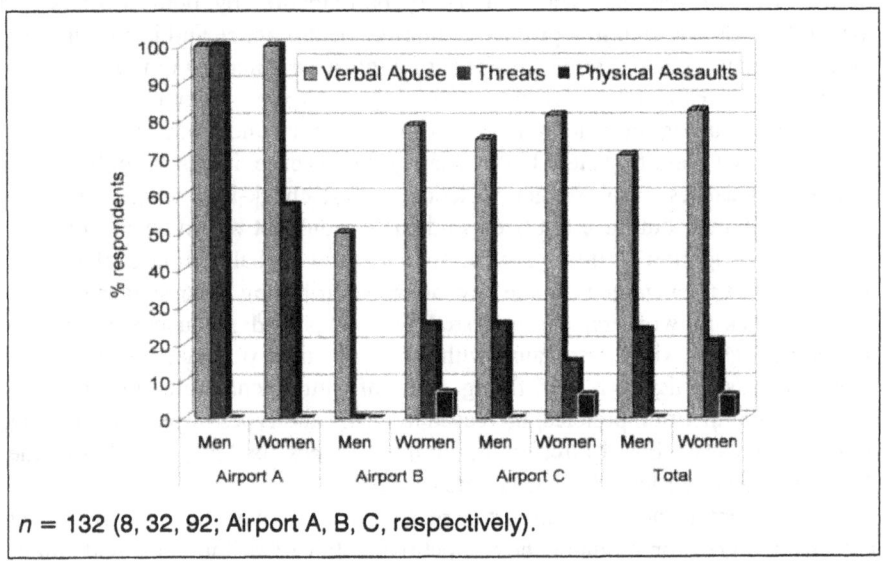

n = 132 (8, 32, 92; Airport A, B, C, respectively).

Figure 3. Experience with violence, by sex
(% respondents having experienced specified type).

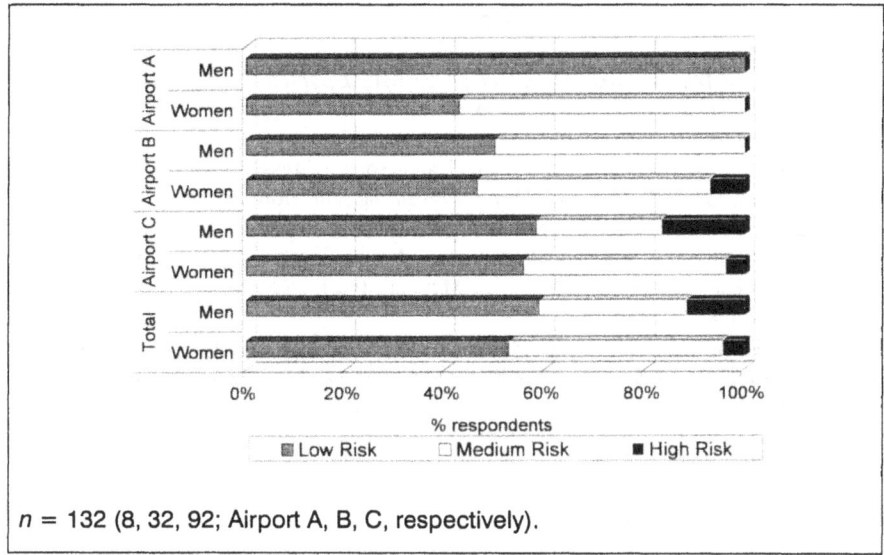

$n = 132$ (8, 32, 92; Airport A, B, C, respectively).

Figure 4. Perceived level of risk related to violence, by sex.

leads to the conclusion that the problem likely knows no social, economic, or cultural boundaries.

In the early 1970s, the adverse effects of new forms of work already were being observed. A 1973 U.S. Government Report *Work in America* reported,

> . . . a growing body of research indicates that, as work problems increase, there may be a consequent decline in physical and mental health, and family stability, community participation and cohesiveness, and "balanced" sociopolitical attitudes, while there is an increase in drug and alcohol addiction, aggression, and delinquency [58].

We go to work to perform a job for which we are paid. Giving away our lives, our health, or our happiness is not part of the employment contract. It is entirely unacceptable that working people must work under threatening or intolerable conditions. People create work conditions. People can make them better. In chapter 6, I present a number of recommendations to prevent airport check-in workers from becoming the victim of work-related violence. Improving work conditions, however, also requires ensuring destructive individuals do not hold managerial positions where they can inflict psychological violence on subordinates. How to identify such people, what to do with them, what kind of harm they can do to their subordinates, and what actions are needed at enterprise or organizational level are the subjects of the next chapter.

ENDNOTES

1. R. Sapolsky, *Why Zebras Don't Get Ulcers: An Updated Guide to Stress, Stress-Related Diseases, and Coping*, W. H. Freeman and Company, New York, p. 309, 1994 and 1998.

2. S. V. Kasl and S. Cobb, Psychological and Social Stresses in the Workplace, in *Occupational Health: Recognizing and Preventing Work-Related Disease*, B. S. Levy and D. W. Wegman (eds.), Little Brown and Company, Boston/Toronto, p. 252, 1983.

3. J. E. McGrath, A Conceptual Formulation for Research on Stress, in *Social and Psychological Factors in Stress*, J. E. McGrath (ed.), Holt, Rinehart, and Winston, New York, pp. 22-40, 1970.

4. S. Aro, Stress, Morbidity, and Health-Related Behavior. A Five-Year Follow-Up Study Among Metal Industry Employees, *Scandinavian Journal of Social Medicine, 9*(Suppl. 25), 1981.

5. R. Kalimo, Stress in Work, *Scandinavian Journal of Work, Environment & Health, 6*(Suppl. 3), 1980.

6. See studies cited earlier in this book related to job strain and cardiovascular disease.

7. R. Karasek and T. Theorell, *Healthy Work: Stress, Productivity and the Reconstruction of Working Life*, Basic Books, New York, 1991.

8. R. Karasek, D. Baker, F. Marxer, A. Ahlbom, and T. Theorell, Job Decision Latitude, Job Demands and Cardiovascular Disease: A Prospective Study of Swedish Men, *American Journal of Public Health, 71*:7, pp. 694-705, 1981.

9. J. V. Johnson, E. M. Hall, D. E. Ford, L. A. Mead, D. M. Levine, N-Y. Wang, and M. J. King, The Psychosocial Work Environment of Physicians, *Journal of Occupational and Environmental Medicine, 37*:9, pp. 1151-1159, 1995.

10. J. V. Johnson and E. M. Hall, Job Strain, Work Place Social Support and Cardiovascular Disease: A Cross-Sectional Study of a Random Sample of the Swedish Working Population, *American Journal of Public Health, 78*:10, pp. 1336-1342, 1988.

11. E. A. Locke, The Nature and Causes of Job Satisfaction, in *Handbook of Industrial and Organizational Psychology*, M. D. Dunnette (ed.), Rand McNally, Chicago, pp. 1297-1349, 1976.

12. Psycho-social factors are understood to be those aspects present in the organization of work and in one's work-related social context which have the potential to create adverse effects on workers' health.

13. Statistics Canada, *Canadian Community Health Survey, Mental Health and Well-Being*, Ottawa, 2002.

14. European Foundation for the Improvement of Living and Working Conditions, *Third European Survey of Working Conditions*, Luxembourg, 2001.

15. European Agency for Safety and Health at Work, *Working on Stress: Prevention of Psychosocial Risks and Stress at Work in Practice*, Bilbao, 2002.

16. Health and Safety Executive, *Occupational Stress Statistics*, Information Sheet 1/03/EMSU, September, 2003.

17. K. Belkic, P. Schnall, P. Landsbergis, and D. Baker, The Workplace and CV Health: Conclusions and Thoughts for a Future Agenda, in *The Workplace and Cardiovascular Disease*, P. Schnall, K. Belkic, P. Landsbergis, and D. Baker (eds.), *Occupational Medicine: State of the Art Reviews, 15*:1, 2000.

18. J. V. Johnson, Occupational Stress, in *Preventing Occupational Disease and Injury* (2nd Edition), J. L. Weeks, B. S. Levy, and G. R. Wagner (eds.), The American Public Health Association, Washington, D.C., 2004.

19. This particular type of job stressor has been described by Arlie Hochschild as "emotional labor," in her studies of airline flight attendants. A. Hochschild, *The Managed Heart: Commercialization of Human Feeling*, University of California Press, San Francisco, 1983.

20. A discussion of these findings is presented in the 1994 National Association of Working Women survey of over 40,000 working women in the United States.

21. N. Sokoloff, *Between Money and Love: The Dialectics of Women's Home and Market Work*, Praeger, New York, p. 211, 1980.

22. K. A. Holte, O. Vasseljen, and R. H. Westgaard, Exploring Perceived Tension as a Response to Psychosocial Work Stress, *Scandinavian Journal of Work, Environment and Health, 29*:2, pp. 124-133, 2003.

23. M. C. Mireya Scarone and L. A. Cedillo, *Thank You For Calling Telmex, May I Help You: Service Interactions and Psychosocial Risk Factors Among Telephone Women Workers in Mexico,* presented at the Fourth International Conference on Work Environment and Cardiovascular Diseases, Newport Beach, California, March 10, 2005.

24. P. Leino, Symptoms of Stress Predict Musculoskeletal Disorders, *Journal of Epidemiology and Community Health, 43,* pp. 293-300, 1989.

25. H. Pikhart, M. Bobak, S. Malyutina, R. Kubinova, R. Topor, H. Sebakova, Y. Nikitin, and M. Marmot, Psychosocial Factors at Work and Depression in Three Countries of Central and Eastern Europe, *Social Science and Medicine, 58*:8, pp. 1475-1482, 2004.

26. J. Siegrist, Adverse Health Effects of High-Effort/Low-Reward Conditions, *Journal of Occupational Health Psychology, 1,* pp. 27-41, 1996.

27. C. Watts and C. Zimmerman, Violence Against Women: Global Scope and Magnitude, *The Lancet, 359,* pp. 1232-1237, 2002.

28. H. Hoel, K. Sparks, and C. L. Cooper, *The Costs of Violence/Stress at Work and the Benefits of a Violence/Stress-Free Working Environment,* International Labor Office, Geneva, 2000.

29. V. Di Martino, H. Hoel, and C. L. Cooper, *Preventing Violence and Harassment in the Workplace*, European Foundation for the Improvement of Living and Working Conditions, Luxembourg, 2003.

30. E. B. Faragher, M. Cass, and C. L. Cooper, The Relationship Between Job Satisfaction and Health: A Meta-Analysis, *Occupational and Environmental Medicine, 62,* pp. 105-112, 2005.

31. A more detailed description of "raising of the bar," for treatment and conditions at work is given in chapter 5.

32. L. B. Anderson and J. E. Smith, *Going Postal: Fact or Fiction?* Presentation at the Society for Human Resource Management National Conference, Minneapolis, Minnesota, pp. 2-9, June 1998.

33. R. Bachman, Violence and Theft in the Workplace, in *National Crime Victimization Survey*, U.S. Department of Justice, Washington, D.C., pp. 1-2, July 1994.

34. D. Anfuso, Deflecting Workplace Violence, *Personnel Journal,* pp. 68, 76, October 1994.

35. A. Pizzino, *Report on CUPE's (Canadian Union of Public Employees) National Health and Safety Survey of Aggression Against Staff*, Ottawa, January, p. 9, 1994.
36. A discussion of the estimated costs of work-related violence can be found in Hoel et al. [28].
37. European Foundation for the Improvement of Living and Working Conditions, *Second European Survey of Working Conditions*, Luxembourg, 1996.
38. C. L. Cooper, P. Liukkonen, and S. Cartwright, *Stress Prevention in the Workplace: Assessing the Costs and Benefits for Organizations*, European Foundation for the Improvement of Living and Working Conditions, Luxembourg, 1996.
39. P. Dorman, *The Economics of Safety, Health and Well-being at Work: An Overview*, International Labor Office, Geneva, 2000.
40. L. Levi and P. Lunde-Jensen, *A Model for Assessing the Costs of Stressors at National Level: Socio-Economic Costs of Stress in Two EU Member States*, European Foundation for the Improvement of Living and Working Conditions, Luxembourg, 1995.
41. K. Chandraiah, D. K. Kenswar, P. L. S. Prasad, and R. N. Chaudhuri, Occupational Stress and Job Satisfaction Among Middle and Senior Managers, *Indian Journal of Clinical Psychology, 23*, pp. 146-155, 1996.
42. M. K. Douglas, A. I. Meleis, and S. M. Paul, Auxiliary Nurses in Mexico: Impact of Multiple Roles on Their Health, in *Health Care for Women International*, pp. 355-367, 1997.
43. For examples, see chapter 7 in *Economic Security for a Better World*, International Labor Office, Geneva 2004.
44. The negative effects on health and well being are immeasurable, and recovery extremely long, when working people are subjected to managers who shout at staff, use physically or verbally threatening behavior, are psychologically abusive, and treat subordinates with disrespect. This issue is discussed in depth in Chapter 5.
45. D. Chappell and V. Di Martino, *Violence at Work*, International Labor Office, Geneva, p. 27, 2000.
46. Discussion of the effects of various forms of harassment on women can be found in P. Crull, Sexual Harassment and Women's Health, in *Double Exposure: Women's Health Hazards on the Job and at Home*, W. Chavkin (ed.), Monthly Review Press, New York, 1984.
47. See in particular the work of K. Rogers and E. K. Kelloway, Violence at Work: Personal and Organizational Outcomes, *Journal of Occupational Health Psychology, 2*, pp. 63-71, 1997; P. Leather, C. Lawrence, D. Beale, and T. Cox, Exposure to Occupational Violence and the Buffering Effects of Intra-Organizational Support, in *Work and Stress, 12*:2, pp. 161-178, 1998; and M. Peze, Corps érotique et corps au travail: Les hommes de metier, in *Travailler, 1*, pp. 1-16, 1999.
48. Evidence from the ILO People's Security Surveys in Brazil, Argentina, and Chile shows that men also report being the victim of harassment in the workplace—mainly physical assault. For a discussion of the findings, see chapter 7 in *Economic Security for a Better World*, ILO, Geneva, 2004.
49. D. N. Castillo, Non-Fatal Violence in the Workplace: Directions for Future Research, in *Trends, Risks and Interventions in Lethal Violence: Proceedings of the Third Annual Spring Symposium of the Homicide Research Working Group*, C. Block and R. Block (eds.), U.S. Department of Justice, Washington, D.C., pp. 225-235, 1995.

50. International Transport Workers' Federation (ITF), Air Rage: The Prevention and Management of Disruptive Passenger Behaviour, in *Safety in Practice, 1,* London, p. 12, May 2000.
51. The extent of violence against cabin crew has been described by the International Transport Workers' Federation [50] and has become the subject of an annual, international media campaign launched by the ITF in 2000. The campaign is to be extended now to include ground staff, including check-in workers.
52. International Air Transport Association, *Guidelines for Handling Disruptive/Unruly Passengers,* Geneva, 1998.
53. P. Crull, Sexual Harassment and Women's Work, in *Double Exposure: Women's Health Hazards on the Job and at Home,* W. Chavkin (ed.), Monthly Review Press, New York, 1984.
54. R. Dickson, T. Cox, P. Leather, D. Beal, and B. Farnsworth, Violence at Work, *Occupational Health Review, 46,* pp. 22-24, 1993.
55. R. Dickson, T. Cox, P. Leather, D. Beal, and B. Farnsworth, Intervention Strategies to Manage Workplace Violence, *Occupational Health Review, 50,* pp. 15-18, 1994.
56. E. Chenier, The Workplace: A Battleground for Violence, *Public Personnel Management, 27*:4, pp. 557-568, 1998.
57. M. Grenier-Peze, Contrainte Par Corps: Le Harcèlement Moral, in *Travail, Genre et Sociétés.* Dossier: Harcèlement et Violence, les Maux du Travail:5, L'Harmattan, Paris, pp. 1-9, 2001.
58. Special Task Force to the Secretary of Health, Education, and Welfare, *Work in America,* Cambridge, Massachusetts, pp. xvi-xvii, 1973, in H. Braverman, *Labor and Monopoly Capital: The Degradation of Work in the Twentieth Century,* Monthly Review Press, New York and London, p. 31, 1974.

CHAPTER 5

Dehumanizing Managers:
Making Workers Sick

Good workers leave bad managers, not bad jobs or bad companies [1].

During conferences, informal discussions, and structured interviews, airport check-in workers, ground staff, and cabin crew—from airports around the world—described abusive treatment that they had been subjected to by supervisors or higher management. The incidents were not isolated events and were reported by workers from various cultures. For some, there were scars. Unacceptable treatment from managers was the reason some workers gave for the intention to leave their jobs.

> The manager shouts at me and I can't take it. It stresses me out and I feel sick and nervous all the time. I am thinking about leaving this job because of her. (Check-in worker)

Destructive or *dehumanizing managers* in the workplace affect workers' physical and mental health and well-being every day, and in powerful ways [2, 3]. The word "manager" here refers equally to supervisors and individuals in any decision-making positions who have authority over others. The management and worker empowerment issues have important and obvious impacts on workers' health. Abusive treatment by destructive or *dehumanizing managers* has been linked to numerous negative consequences, including MSDs, depression, anxiety, cardiovascular disease, myocardial infarction, ulcers, cancer, hypertension, death from stress-induced diseases including suicide, and a range of other adverse health outcomes. A study of angina pectoris in Israel found that men who reported problems with supervisors were at greater risk for recurring angina [4]. Associated conditions included workers who collapsed in their offices or outside the workplace; attacks of amnesia from working under destructive managers; gastro-intestinal disorders; herpes; alcohol, drug, and tobacco abuse as coping behaviors [5, 6]; and the list goes on [7].

No less significant are the major cost factors for workers, employers, families, and insurance companies that result from the psychological distress-induced health problems that many workers suffer as the subordinates of emotionally destructive managers.

The point in discussing this issue is to educate working people, organizations and companies about the characteristics of these bosses so they understand how dangerous abusive managers are, how to recognize them and their patterns of behavior, and learn what they can do about the abuse.

The type of destructive people described here can be found also in non-managerial positions. Destructive individuals inhabit all levels of a hierarchy. This chapter, however, will focus only on destructive *managers* and the impacts they have on the well-being of their subordinates. I focus in particular on managers because of the force they can wield over subordinates, the potential for them to abuse power, the tremendously negative effects they can cause on the health of subordinates, and the unequal power dichotomy that still exists between managers and subordinates. Managers can get away with much more than subordinates in most workplaces and leave subordinates in a position of disadvantage. Indeed, destructive managers constitute an important form of *excess baggage,* so excessive in fact that no payment can ever equal the damage done to those having to bear their weight.

Dehumanizing managers are Dr. Jekyll and Mr. Hyde personalities. This view is one way to think of them that may feel familiar. One side of these people appears as the "good guy," the good citizen in the workplace, someone who wants to help people, do right by their subordinates, and so forth. They may tell their subordinates that they stick up for them, that they go to bat for "their people." But in reality, they do nothing of the sort. It is all talk, part of their seduction and manipulation act. The Mr. Hyde side of their personality is insidious, destructive, and dangerous. And it dominates more than the Dr. Jekyl "good guy" side. World wide the vast majority of managers are men. This fact very likely explains the most common workplace harassment cases: male superiors who are abusive to women subordinates (through the use of various forms of abuse and harassment).

PRECIOUS LITTLE RESEARCH

A number of people (all in managerial positions) have asked that this chapter include a discussion of psychological distress and strain among managers. It is a valid area to explore, but it will not be addressed here. There is plenty of research on stress and strain among managers, evidenced by a relatively small list of research studies referenced in this chapter [8]. Our study of unionized check-in workers did not include managers.

What is not available, however, are scores of (or even simply more than a few) research results on the negative impacts on workers' physical and psychological health from the emotional abuse from working for destructive managers. Even less information on this subject is written from a worker empowerment perspective. There is precious little discussion about the obvious association between destructive managers and ever-increasing work-related psychological distress and physical harm, with their associated costs. The information gap is surprising, given that working people spend most of their waking hours at work, in the sphere of their supervisor or manager. Little of the research literature, or even trade union and international organization documents on workplace violence, include discussions about psychological violence in the workplace, and few recognize moral and sexual harassment as forms of violence and abuse.

The discussion about psychological violence in the workplace has been initiated, but unfortunately only by scraping the surface, often inappropriately, and by understating the problem [9]. Psychological violence in the workplace is more common than we realize. At its extreme, psychological violence in the workplace aims at destroying the professional skills of an employee and breaking her down mentally, as a human being. The willingness to recognize the cruelty of psychological violence in the workplace as well as widespread measures to eliminate it does not seem to exist. Occupational health and safety institutions and governmental bodies are not up-to-date on the issue, or on the legislative aspects. How can a labor inspector, therefore, be capable of concluding that a workplace is fit for work? An important knowledge gap concerns the over-arching behavior patterns typically seen in the individuals committing harassment in the workplace or acting abusively toward others. Those committing psychological violence in the workplace follow the same patterns, as if out of a handbook. Their patterned behavior, however, facilitates the identification of abusers at work.

There is an important research and discussion gap on this critical link in the literature on MSDs, work-related stress, workers' social and economic insecurities, social policy, workplace violence, or any health outcomes literature. Mostly, the impacts of destructive management on workers are discussed in books and articles on psychology, psychiatry, and workplace harassment. The main argument defending the lack of discussion in health, social, economic, or labor-related publications is that a large body of research is needed to make clear scientific associations. This argument often is made by researchers whose sole aim is to publish their articles in scientific journals.

While the knowledge gaps continue to be glaring, more working people are suffering, getting hurt, and falling ill, and the costs to employers, workers, their families, insurance companies, and society continue to rise. Generalized abusive experiences in the workplace have been shown to be more prevalent than sexual harassment. Both sexual harassment and generalized abuse are significantly related to symptomatic distress, alcohol and drug use and abuse. Women workers more often are the victims of workplace abuse and its serious health effects than men [10]. Any discussion about the impact of management practices on workers' health should include discussion about destructive and abusive [11] managers in the workplace. While management practices in general, poor work conditions, the increasing speed of production, lack of worker involvement, gender and sex discrimination, employment insecurity, and so forth are important factors that contribute to workers' distress, the potential for negative psychological and physical health effects caused by destructive managers cannot be overstated.

> . . . emotional abuse is hard to describe, but its nature is open to systematic investigation, although the studies to date are limited methodologically and conceptually. . . . Most attention has focused on the more extreme forms of physical violence such as homicide, as well as, to a lesser degree, sexual and racial harassment in the workplace. However . . . emotional abuse is more frequent [in the workplace], has similar effects as physical, sexual, and racial violence, and yet appears to be more socially acceptable. Emotional abuse refers to the hostile verbal and non-verbal behaviors . . . directed at gaining compliance from others. Examples of these behaviors include yelling, screaming, using derogatory names, the silent treatment, withholding necessary information, aggressive eye contact, negative rumors, explosive outbursts of anger, and ridiculing someone in front of others. . . . In addition, factors that have not yet been considered in any depth in the workplace abuse literature include the target's response to the abuse behaviors . . . [12].

Additionally, and no less important, other than books and articles on psychological and sexual harassment and mobbing and bullying in the workplace, there is a glaring lack of research discussing the kinds of workplaces that *allow* and *enable* a culture where harassment in its various forms takes place. Findings from a study by Shahtahmasebi [13] demonstrated that bullying (whether due to isolated individuals, competition, rivalry, power, or pure meanness) can thrive only in a bullying management culture, and that the consequences include impacts on the work environment, damage to individuals' health, and economic and financial loss for the enterprise or organization [13].

> Business, as well as the media, has tended to focus on sexual harassment, which is only one aspect of harassment in a larger sense. This psychological war in the workplace consists of two elements:
>
> • Abuse of power: often quickly revealed and not accepted by the employees
> • Emotional manipulation: more insidious and more destructive from the beginning [14].

Bullying as one particular type of abuse in the workplace has been associated with an increase in cardiovascular disease, depression, and other mental health problems, and generalized distress [15]. Higher levels of general strain and mental distress have been found among workers who have been the target of bullying as well as the observers, compared with workers in workplaces with no bullying [16]. A body of research exists on violence in the health sector in particular, with some focus on the prevalence and impacts of bullying in the workplace. Bullying, it has been suggested, is largely a problem of misinformed management [17-20]. Research findings point to an important incidence of bullying in workplaces [21]. Such findings need to be linked with studies demonstrating certain types of management cultures that act as enabling environments for abuse toward working people.

Reasons for the lack of discussion about the important and growing problem of the impact of destructive managers on the health of their subordinates include:

- Many researchers do not ask the right questions in order to know that destructive management is a critical growing cause of work-related psychological distress and ill health. They often fail to ask the right questions because they do not conduct research in collaboration with workers, who could describe the causal relationship;
- Many occupational health and safety researchers look for factors that are relatively easy to measure and quantify, and as such may overlook material that emerges from qualitative research;
- Until recently, the subject of emotional abuse in the workplace through psychological violence has been largely neglected;
- Work-related stress research is relatively new and not well supported (because much occupational health research is controlled by management and governments and the findings of what research has been done show that radical changes are needed in the way work is presently carried out);
- It is easier to investigate and measure use of alcohol, tobacco, and drugs as indicators of psychological distress among working people (these often are used as a means of coping with intolerable situations of chronic psychological harassment in the workplace) than to design studies to demonstrate the negative impacts from working under destructive managers, compared with the outcomes of workers under emotionally healthy managers.

(Additional reasons are provided in the Endnotes to this chapter [22].)

RAISE THE BAR AND NIP IT IN THE BUD

> They told me to pick up any bag, no matter how heavy it is. I won't do it. If a bag is very heavy, I tell the passenger to pick it up and bring it over himself. The supervisor told me I have to do it, but I don't care what he says. I'm not gonna' hurt my back for anyone. He kept telling me that and got nasty about it. I don't care though. (Check-in worker, Airport C)

Questions have emerged from discussions with check-in workers. 1) Why will one twenty-year-old female airport check-in worker tell a male passenger bringing in heavy bags to lift and carry the bags himself, while no other check-in workers will? 2) Why will one worker tell her boss to take a hike, and walk out of the boss's office the moment the boss raises his voice, while others will allow the boss, or a client or passenger, to shout like a two-year-old having a temper tantrum? 3) Why is that "nip-it-in-the-bud" [23] approach instinctive for some and not for others (can it work for everyone, everywhere; does it depend on relative rank, personality, pulchritude or age)? 4) How does one learn to set their bar [24] sufficiently high, so that they immediately recognize as unacceptable any behavior or conditions falling below their bar?

These questions hold importance for identifying new ways of translating the concepts of worker empowerment into practice in today's work climate. The questions became important in determining whether the tools found in those who know how to "nip it in the bud" could be identified and taught in the workplace to non-bud-nippers. If so, then teaching non-bud-nippers to raise their bar, demand respect, and to stand up for oneself in the face of unacceptable treatment or work conditions, might help create critical masses ready to exercise their voice, refusing unacceptable treatment or conditions. These questions hold increasing importance the more one notices that it is only to a certain degree that working people can rely solely on "systems" and policies to protect them. At a certain point, individuals have to recognize what is unacceptable and state clearly that they will not accept it. That means taking responsibility, which in most workplaces most people do not do instinctively.

There is a great deal of talk about the need for laws and regulations, and good practices that flow from top to bottom in a workplace. Indeed, they are necessary, but by themselves never enough. And should they be? Implicit in government by the people and for the people is that we, the people, ensure governance belongs to us, which means that it is largely up to working people to ensure that laws and policies are brought from the level of state governance to the local level workplace. Practices flowing from the bottom upwards are needed while top-down practices are sought or implemented [25]. The use of both individual and collective "voice" is, therefore, essential in ensuring that one's rights are respected at work. Even then, there is no guarantee. Broad-based teaching of bud-nipping tools might increase awareness of how to take more responsibility for oneself in the workplace and demonstrate where one's bar should be set— below which any behavior, treatment, or conditions are considered unacceptable.

Where to set the bar for self-preservation can be taught and can be learned. There is no single package that can teach these tools to everyone, everywhere. Some training programs exist, often called "assertiveness training." But to be worthwhile, they need to help you critically analyze why your bar has been set too low, behavioral dimensions to watch out for, as well as provide tools for asserting yourself in the face of unacceptable treatment or conditions. Learning assertiveness tools without the critical self-analysis is only half the battle. And one has to be ready to take the steps.

Having an effective union in the workplace can make a significant difference. It can provide support in the face of unacceptable treatment or conditions. An effective union can determine jointly with management what constitutes unacceptable treatment and conditions, and what to do about them. And it can take the lead to change the cultural hegemony so that workers are put first—not managers and not capital. After all, it is workers who produce the value that we enjoy. Unions must tackle these challenges at the level of the individual workplace and the level of society (described under the subheading *Strategies for Prevention* later in this chapter).

RECOGNIZING THE UNACCEPTABLE

Destructive managers can appear at any level of a hierarchy, from which they abuse their position of authority and treat those who report to them in a *dehumanizing* manner. The effects on workers are the same no matter what the level of hierarchy of the manager. Abusing authority and *dehumanizing* others in the workplace is unacceptable at any level. It is well demonstrated, however, that the higher up the hierarchy, the more insidious the dehumanization and destruction by the manager. In manufacturing and production sectors, psychological violence by managers is more direct, verbal or physical. In white-collar workplaces, *dehumanizing managers* often re-write what their subordinates produce. They only accept output that is entirely in their own words and that reflects only their ideas. The more one rises in the hierarchy or on the socio-cultural ladder, the more the aggressions are sophisticated, perverse, and difficult to repair [26].

Above all, starting to think about where one's bar is set is a good first step. Taking responsibility for what one is willing to accept in the workplace is a good next step. And taking action is an excellent third step. They are steps to create positive change and help others learn along the way. Learning to "nip it in the bud" can be seen as workplace-based prevention and health promotion through self-empowerment and the use of voice.

Little in the work environment encourages questioning things "as they are." Management by command and control thrives on worker passivity. Ask the proverbial "why" at work and typically one is put in one's place. Only the brave question the way things are done. Such people usually look for ways to increase simplicity and efficiency and want transparency and equality in the workplace; they do not question just for the sake of exposing. Considering the dominant culture of management by control, why do managers wonder why they cannot get "their people" to produce enough or work harder? Managers would do well to look at what the mirror reflects.

Authority is, of course, necessary for social order and social cohesion. What is unacceptable, however, is the abuse of a position of authority, particularly decision-making authority, as if it comes with an ordained right to exercise control over others, to treat others with disrespect, be destructive toward others, or to do whatever the manager wants irrespective of the consequences for other people. If a working person feels her dignity or psychological integrity damaged because of the ongoing hostility of a boss over a long period of time, it is most probably a matter of emotional abuse. People in positions of authority in the workplace may interpret their position as one of superiority, of entitlement, believing they are better than others and that those below them are not worthy of the respect that they, themselves, expect and demand.

Such people rarely consult anyone in making decisions in the workplace, even though the decisions they make impact many people. It is as if working people around such managers do not exist and do not count. In short, such managers usually attribute far more importance to themselves than they deserve. After all, they are employed to do a job, just like everyone else with whom they work.

Key roles for any manager are to ensure that workers have the tools they need to perform their job; to create and provide an enabling environment conducive to performing the job; to support, motivate, encourage, and show appreciation and validation for people's work; and to further the self-affirmation of every worker. These are fundamental human needs. Fulfilling these needs costs nothing and is paid back to an employer many times over. A manager is not employed to build obstacles to the performance of one's job or to create a disabling, disempowering, or dehumanizing environment where the performance of one's job is not facilitated, or where one's work is not recognized, validated, or appreciated.

People in positions of authority are at high risk of losing their powers of perception. There is, therefore, an indisputable need to have checks and balances for those in managerial and other positions of decision-making in any workplace. This is a key means of keeping abreast and understanding whether practices being applied by management—such as quick turnaround practices in airports—result in detrimental effects to workers and to the enterprise. Allowing managers to have total control over what they do is not only risky business-wise, but it creates a high risk for abuse in the workplace. "In the workplace, resources are to be exploited; people are to be respected" [26, p. 248].

One should not be fooled if authoritative managers occasionally ask for input from subordinates. Usually, they want merely superficial consultation. Patterns of behavior tell much more than one-off (superficial) consultations. Are workers regularly and consistently consulted on all or at least critical aspects of decision-making? Are the consultations about important issues or only insignificant matters, tokenism, meant to make employees feel "involved?" Is anyone else's suggestion or vision ever implemented, enacted, shared with others, or even seriously considered? Nine (or perhaps ten) times out of ten, an authoritative figure in a management position will act only on what he alone thinks, much to the chagrin, dismay, and disillusionment of those around him, and irrespective of the consequences for others.

Human beings need to be treated with dignity, decency, and respect at work, and work cannot be decent without respect. For workers at any level of a hierarchy not to be involved in and able to influence decision-making is to treat them as children and is by definition disrespectful of workers' knowledge, competencies, and experience. It is one of the best-known ways to dumb and numb the masses. Frederick Taylor would be proud, since he firmly believed in keeping workers dumbed and numbed through authoritative management practices. The use of such practices today is, in reality, a modern-day version of Taylorism. "There is no way to treat employees as responsible and honest adults unless you let them know and influence what is going on around them" [27, p. 117].

Probably the main reason for the lack of discussion about the obvious, direct, and serious negative effects on workers' health from destructive management and the important associated costs is the lack of worker involvement in workplace decision-making. Modern-day management does not want workers involved in organizational decision-making. Doing it by management alone is easier and means there will be no challenge to absolute rule. Managers tell workers that they should be happy just to have a job and remind them that they are replaceable at any time. The growing insecurities today mean that increasingly workers keep quiet and are frightened to demand that their rights be respected. Many managers, believing that no one can stop them, take full advantage of this insecurity and use it to squash and destroy workers.

There are, of course, many good, effective, and humane managers in workplaces all over the world who treat people with respect and dignity. Generally, everybody in a company or organization wants to work with them. Companies and organizations with good managers have low rates of worker turnover, which is excellent for motivation, productivity, and profit-generation. Most managers are not abusive, dangerous people, but there are plenty of managers everywhere who are abusive, perverted people, who treat subordinates with no respect or dignity. They are the ones to look out for and to avoid at all costs. One reliable indicator of a destructive manager is a high rate of worker turnover among his subordinates.

RECOGNIZING DESTRUCTIVE MANAGERS

Many books and studies address various forms of authoritative personalities, not all of which are necessarily destructive (although one could debate that position). Without going into a broad psychological analysis of authoritative personalities, it is important to emphasize one particularly insidious, destructive, and virulent type that goes well beyond simply being authoritative and can be identified increasingly among managers today. In fact, it is a personality disorder (sometimes called a personality deficit [28]), whose causal factors (usually in childhood) lead this type of person as an adult to not trust anyone. People so afflicted feel that society and everyone in it "owes" them, although in reality no amount of "payback" will suffice. There are hypotheses about genetic origins, but the data are insufficient to support this theory. The scientific term for emotionally abusive managers or supervisors is *narcissistic pervert* [29, 30].

> Narcissistic personality traits are common enough (egocentricity, need for admiration, intolerance of criticism), but they are not necessarily pathological . . . a perverse narcissist only functions by gratifying his/her destructive impulses [14, p. 123].

The destructive managers described in this chapter are *dehumanizing managers*. By definition, such managers are abusive narcissists. They dehumanize their subordinates. *Dehumanizing managers* do not see others as worthy of respect and so they do not treat anyone with respect. They establish relations with others based on force, mistrust, and manipulation [26, p. 345].

Here are glaring signs that reveal *dehumanizing managers* in the first or early encounter:

1. Has a tendency to humiliate, criticize, and belittle others;
2. Systematically puts blame or responsibility on others for whatever happens;
3. Egocentric;
4. Jealous;
5. Cannot stand criticism or denies evidence;
6. Lies;
7. Questions the qualities, competencies, personalities of others, devalues and judges;
8. Alters and interprets so that it suits his needs;
9. Often only at the last minute makes requests or gives orders to make other people take action;
10. Changes the subject in the course of a conversation;
11. Perfectly efficient at achieving own objectives but at the expense of others;
12. There is a sadistic side to such people;
13. Creates conflict and suspicion among co-workers to maintain total control and to provoke ruptures between co-workers;
14. Their spoken and written words seem logical and coherent while their attitudes, actions, or behavior correspond to the opposite scheme;
15. Does not communicate clearly demands, needs feelings, opinions;
16. Displays arrogant behavior;
17. Has haughty body language;
18. Has a tendency to exaggerate, uses small, unnecessary lies;
19. Finds everyone else guilty;
20. Has a tendency to fantasize about unlimited success;
21. Brags incessantly, to ignore you, not to listen;

22. Changes opinions, actions, feelings according to the people or the situation;
23. Makes others think that they should be perfect, that they should never change their mind, that they should respond immediately to his demands and questions;
24. Ignores the requests of others;
25. Uses moral principles to appease his needs;
26. Constantly the object of discussion among people who know him, even if he is not present;
27. Often answers in a way that is not clear or straightforward;
28. Does not recognize the rights, needs, or desires of others;
29. Makes promises which are incommensurate either with the event or with his ability to fulfil them;
30. Transmits messages by someone else or by an intermediary [31].

The list presents a selection of prominent characteristics that are common in *dehumanizing managers*, but one needs to identify only five from the list to determine that you work under a genuine manipulator, someone psychologically unhealthy and destructive in nature.

The clinical experience and research of Marie-France Hirigoyen demonstrate that the dysfunctional behavior of a *dehumanizing manager* displays at least five of the following six "symptoms":

1. A grandiose sense of his importance;
2. Fantasies of unlimited success/power;
3. Thoughts of being unique and "special";
4. Belief that everything is owed to him;
5. Exploitation of others in relationships;
6. Lack of empathy [14, p. 125].

In the minds of such managers, no one's work is good, only their own has value, and as such, they will criticize, put down, and thoroughly devalue everyone else's work. This attitude has a powerful de-motivating effect on workers. Indeed, in many workplaces after a short time working under a *dehumanizing manager*, workers switch—consciously or subconsciously—into "minimum output mode," knowing that nothing they do will be praised, appreciated, or valued in any way. Work is merely dictated to subordinates and never decided in conjunction with the workers. *Dehumanizing managers* often complain to their subordinates that they have to do everything themselves because they cannot rely on their subordinates to work properly. Such people present themselves to their subordinates as the victim of their incompetent or lazy underlings, who thus create extra work for the boss. Organizations and enterprises that accept the practices of such destructive individuals are unhealthy places to work.

No *dehumanizing manager* will ever recognize himself in any description of his personality disorder. They never do.

ABUSIVENESS FASCINATES

Other than himself, no one else really exists for a *dehumanizing manager*. Other people are tools for their use or their own objectives, pawns in their chess game. The *dehumanizing manager* recruits people in the workplace to do their bidding and be their servants. They control and use workers, staff, and colleagues to tempt, convince, communicate, perform, threaten, and otherwise manipulate for them. They control often gullible and unsuspecting co-workers, subordinating them with their seductions and manipulations. Other people around them at work, particularly those in a subordinate position, are there only to serve this kind of destructive boss. Workers' schedules, time, work demands, and priorities, do not matter for such managers. The *dehumanizing manager* never questions himself and experiences no regret for any of his actions. And they dump their props (meaning those who fed their ego) unceremoniously, in frightening abruptness, when the job is done, usually causing tremendous and typically irreversible emotional hurt.

Dehumanizing managers use seduction, displayed through their considerable charm, persuasiveness, manipulativeness, and thespian skills. The more intelligent, the more they are talented actors. They may act and speak with great bravado, a seemingly natural born leader, the center of attention. All available tools are fair game, but seduction, in one form or another, including sexual and intellectual seduction, is used as a tool to hook their "prey." They sniff out innocent, well-meaning people who will make useful tools for their purposes and use seduction to capture the unsuspecting victims in their web. Seduction can be exerted through a winning smile, constant charms, invitations, outreach to a particular individual to make that person feel important, compliments, particular looks, gestures, comments that seem particularly friendly and well intentioned, sexual overtures, impressing with their intelligence and accomplishments, and so on.

> [Dehumanizing managers] . . . often choose as their victims people full of energy and love of life . . . [the dehumanizer's] deficiencies are shown up by the desires and vitality of the other. They need to subjugate their victim to secure their own strength [14, p. 131].

Many people, male and female, are weak when they face the seductive powers of *dehumanizing managers*, believing at first (and often forever) that they are just interesting and charming people, even brilliant when they are very intelligent and highly educated. If you do not have the skills to recognize their game of seduction and if you do not show them that their seduction will not work with you, then you will be targeted immediately to become another pawn in their game.

How can one distinguish a *dehumanizing manager* from someone who is genuinely nice, friendly or charming? It is necessary to "screen" and to keep enough of a distance until you have sufficiently screened. Some people have built-in radar to identify (and protect themselves against) destructive individuals. Others have to learn how to screen, learn what to look out for, and learn how to resist and be wary of destructive personalities in the workplace. Learning such tools is critically important in order to avoid the pitfalls that can lead to extreme work-related distress with potentially serious negative health outcomes.

The *dehumanizing manager* needs subordinates and co-workers to adore him, to be awed, to admire his deeds and capabilities, to learn to blindly trust and obey him, to applaud, and ultimately surrender to his charisma and become sub-merged in his follies. *Dehumanizing managers* use others to fulfil their ego needs. They are individuals who are immeasurably preoccupied by their ego. They need to succeed and be admired [26, p. 341]. They often set subordinates against each other. They lie about everything if it is in their interest, yet may give lessons in honesty and integrity to others. They are not self-aware and their cognitions and emotions are distorted.

Abusiveness fascinates, seduces, and terrifies. Often abusive individuals are imagined as endowed with a superior strength that will always make them winners. They do, in fact, know how to naturally manipulate and this appears to give them the upper hand whether in business or politics. Fear makes people instinc-tively gravitate toward them rather than away from them. The most admired individuals are those who enjoy themselves the most and suffer the least. Their victims, who often seem weak and dense, are not taken seriously, and under the guise of respecting another's freedom, we become blind to potentially serious situations.

In fact, this tolerance prevents us from interfering in the actions and opinions of others, even when they seem out of line or morally reprehensible. We also weirdly indulge the lies and "spin" of those in power. The end justifies the means. To what degree is this acceptable? Don't we, out of indifference, risk becoming accomplices in this process by losing our principles and sense of limits [14]?

In reality, as soon as a *dehumanizing manager* is not at his best, most brilliant, most performing level, they have the feeling that they do not meet what is expected of them. While this characteristic explains in part the destructive behavior of such people, it is no excuse for their dehumanizing treatment toward others in the workplace [26, p. 341].

Dehumanizing managers lack empathy. To a *dehumanizing manager*, he is never wrong—everyone else is always wrong. No matter how destructive his behavior is toward others, he never sees himself as responsible or wrong in his actions. He offers plausible renditions of any and all events and interpret them to his favor [31, p. 41]. The *dehumanizing manager* often presents himself as a victim, when in reality, it is he who does the victimizing. The fact that the *dehumanizing manager* feels that everyone "owes" him, and that he sees himself as a victim, justifies—to him—whatever he does. *Dehumanizing managers* need to fully control everyone around them, including the emotions of others.

Listening to the way a *dehumanizing manager* talks, it becomes apparent that no matter what happens in the workplace, they never sees themselves as responsible in any way, despite their managerial position, which entails at least some degree of decision-making authority. They never acknowledge the slightest mistake or blame on their part. Invalidating others is a way of not seeing themselves [14, p. 13]. Everything will be explained as "objective events" beyond their control, as someone else's fault, or due to someone else's incompetence. They describe events as the system working against them, so that others see them as victimized by the system and therefore entitled to more than others. *Dehumanizing managers* are paranoids and cowards [31].

DEHUMANIZERS HARASS

Dehumanizers in managerial positions often commit moral (psychological) or sexual harassment in the workplace, and it is quite likely that those who commit harassment in the workplace are always *dehumanizers*. The victims of the harassment often arrive at the verge of a nervous breakdown: harassed, a physical and emotional mess, irritable, impatient, abrasive, and hysterical. The *dehumanizing manager* is completely insensitive to what happens to the subordinate. Confronted with this contrast between a polished, self-controlled, and suave abuser and their harried casualties, it is easy to conclude that the real victim is the abuser or that both parties abuse each other equally. The prey's acts of self-defense, assertiveness, or insistence on her rights are interpreted as aggression, instability, or a mental health problem. This scenario has been well described and substantiated in publications on harassment in the workplace [14, 26, 31, 32]. The *dehumanizing manager* often says that he, in fact, was the real victim of harassment rather than the person who claimed to be the victim. He turns the picture around, offers a plausible rendition of events, and interprets them in his own favor. Be very sure of one thing—a *dehumanizing manager* is never a victim, in spite of their very convincing arguments that they are [26].

Psychological harassment is characterized above all by repetition of the behaviors. It is the attitudes, words, and behaviors of managers which taken separately or individually may seem harmless but whose repetition and systematic use makes them destructive. Individuals who repeat such behaviors are by definition *dehumanizers*. In general, the term psychological harassment is inappropriate when describing a one-time attacking attitude from a manager, that is, a punctual event that happened once, even if it had serious consequences for the worker. A one-time verbal attack (unless it was preceded by multiple small attacks) is an act of violence, but it does not constitute psychological harassment; the repetitive and systematic nature of the act was not present. However, if repeated put-downs are involved, then the treatment constitutes psychological harassment [26, p. 37].

> I had a really abusive boss once at an airport I worked at. He was horrible, a real harasser. I fell apart in the end and quit. He told me to work faster all the time. He lied to the management a lot but you know, the managers always stick together. He was really nasty. He would always come and say things to me like I was not doing it right or I wasn't good at my job. He was much worse than the bad passengers. I was too scared to say anything to anyone. I left in the end but I was really depressed after that, and pretty nervous for a long time. Why did I have to get that boss? (Check-in worker)

It is often difficult to distinguish psychological harassment from other poor work conditions, and it is at this juncture that the concept of "intent" takes on importance [26, p. 40]. Being forced to work in unacceptable work conditions is not psychological harassment unless the conditions are forced on a particular worker or the conditions are purposefully meant to demoralize and discourage the worker. Overwork and an increasing or unrealistic workload do not constitute psychological harassment unless they are, consciously or unconsciously, directed at one worker in order to break that individual. The difficulty arises when, in the case of psychological harassment, the degradation in work conditions is progressive and the worker is unable to judge at what moment her work conditions became abnormal. If a group of workers in the same workplace notice that their work conditions are getting worse, collective action is possible. However, an individual worker alone in an organizational structure can rely only on her subjective feelings to make a complaint. Most of the time, psychological harassment is so subtle that the degradation of an individual's work conditions is not at all obvious in relation to the co-workers' situations [26, p. 41].

Some individuals are indifferent to criticism, threats, or judgments made of them. They do not feel guilty or responsible for what happens around them. In general, such people are indifferent to the opinions of others and do not even realize that their *dehumanizing manager* is treating them in an unacceptable manner. These individuals are extremely resilient. While their resiliency protects them, it also means that most of the time they do not know how to recognize a manipulator or *dehumanizing manager* and they therefore negate the existence of a real problem when it exists around them. These "indifferents" do not realize or understand the suffering of a worker victimized by a *dehumanizing manager*. However, the vast majority of us are vulnerable to the manipulations of *dehumanizing managers* [31, pp. 180-181].

The *dehumanizing manager* has to dominate his subordinates at any cost. One can have a strong personality without dominating others. In fact, dominance suggests that there is little personal power in the person who needs to dominate others. The *dehumanizing manager* destabilizes subordinates so they are incapable of responding. In this way, the destructive manager locks his subordinates into permanent submission. Such behavior constitutes psychological harassment. Tremendous inner tension and a high degree of chronic psychological distress are outcomes of working for a *dehumanizing manager*.

Because the *dehumanizing manager* sees himself as never responsible, never guilty, he will do whatever it takes to convince others of it. He believes that the psychological violence he inflicts is deserved (frightening, but important). The harassment he inflicts on one subordinate is always a pattern of behavior, that is, the manager did it before and always will do it again. The psychological harassment inflicted by *dehumanizing managers* on subordinates can destroy a person's life and even lead to suicide [33]. It can take years for the victim to recover—if they ever do. This fact is a strong justification for removing *dehumanizers* from managerial positions and ensuring that others with the characteristics never reach such positions in the first place.

With *dehumanizing managers*, there is no such thing as "fair game"—they never play by the rules, do not respect rules, and are unable to acknowledge and abide by personal boundaries set by others. *Dehumanizing managers* co-opt the system, whatever the system, including their own direct bosses, CEOs, mediators, lawyers, judges, and police officers. In cases of workplace harassment, without people in both human resource and legal positions skilled at recognizing these destructive people and their patterns of behavior, victims of harassment often become inadvertently sacrificed by the system and re-victimized all over again.

This scenario is not at all uncommon [34]. Particularly in cases of harassment at work, *dehumanizing managers* will use their powers of seduction, intelligence, thespian skills, and manipulativeness even to co-opt their defense attorneys into presenting the victim as the actual perpetrator in a court of law. This is precisely why workers find it so difficult to win a psychological or sexual harassment case in court against a *dehumanizer*—because the perpetrator always lies. Many such instances of this destructive and pathetic scenario are described in literature related to harassment in the workplace [14]. Once submerged in the *dehumanizing manager's* web, the only way to get out of doing their calling is to stop working for them, unless there are enough people who recognize the personality type and succeed in getting the individual removed from managerial functions.

DEHUMANIZERS DESTROY WORK CONDITIONS AND WORKER HEALTH

Dehumanizers often are found in positions of decision-making authority but also can be spotted at any level in the workplace. There are degrees of *dehumanizing managers*; some are more talented than others in their ability to destroy and therefore are more dangerous. There are your average little *dehumanizing managers* in the workplace [35] and then there are the brilliant *dehumanizing managers*. Both kinds can make it into management positions. However, the more intelligent the *dehumanizing manager* the more destructive he is because the better he plays the game. Chess players have to see several moves ahead. Those *dehumanizing managers* who can see several moves ahead are the most dangerous. Many people can be fooled into thinking these managers are simply excellent strategists or highly logical thinkers but falling into this trap is a mistake and a misjudgement. They may indeed be talented strategists or even brilliant in a particular field of work. However, left un-checked in managerial positions where they make decisions about other people, their predictable destructive behavior patterns continue. Is the damage they do to others around them and to the company that employs them a fair trade-off for their talents in one or another area of work? Obviously, there is no justification for sacrificing workers for the benefits obtained from such people. There is, however, justification for removing them from managerial positions and allowing them to continue their work for the employer in a different capacity where they cannot have a negative impact on the lives of people around them.

In the workplace, it is critically important to know how to recognize *dehumanizing managers*, particularly because they are insatiable and vindictive. *Dehumanizing managers* feel deprived and unfairly treated, and always present themselves in this light, demanding that they be "protected" by their employer and that their own rights come before all others'. Because *dehumanizing managers* are possessive, they make it difficult to get away. They demonstrate a panoply of indiscriminate negative emotions, including rage and envy. They may shout at subordinates. They have a propensity for reckless behavior. Since other people do not exist for them, they do not care how they treat others or how others feel. Such people categorically should not be in managerial positions [32].

Paulo Freire wrote about "*dehumanizers*" in the late 1960s as a way to describe oppressors [36]. Freire described the treatment of oppressors over peasants in relation to class struggle but his description bears an uncanny similarity to that of the *dehumanizing manager* found in today's workplaces. Freire described those who oppress: ". . . such people [the oppressors] dehumanize others and violate their rights" [36, p. 38].

This tendency of the oppressor consciousness to "in-animate" everything and everyone it encounters, in its eagerness to possess, unquestionably corresponds with a tendency to sadism [36, p. 41].

The oppressed, as objects, as "things," have no purposes except those their oppressors prescribe for them [36, p. 42].

Any situation in which "A" objectively exploits "B" or hinders his or her pursuit of self-affirmation as a responsible person is one of oppression.

> Such a situation in itself constitutes violence, even when sweetened by false generosity, because it interferes with the individual's ontological and historical vocation to be more fully human [36, p. 37].

Similar to Freire, Marie-France Hirigoyen also has described the hatred and sadism demonstrated by *dehumanizers* [14, p. 93].

Dehumanizing managers cannot stand criticism, challenge, contradiction, or disagreement, and they will not put up with it at any cost from subordinates or superiors. It is extremely important for them to project a particular image about themselves, and they perceive any affront to that image as a "punishable" offense. For example, criticizing, challenging, or even questioning by a subordinate at work can be construed as good cause to belittle, put down, or even fire the subordinate. However, the reason for a putdown or firing will be presented to observers as something entirely justifiable. One can see workers who have excellent qualifications and competencies, and who may have been nice co-workers, fired simply because they questioned their *dehumanizing manager*. They never knew what hit them.

> Some of the supervisors are real bastards. They shout at you and if you say anything back you really get hell. They always think they're right but hey—it's amazing how often they don't know what they are talking about! If you say something that's wrong, they might give you the worst shifts. Then if you complain, it just gets worse. They can be really mean sometimes. (Check-in worker)

Conversely, and equally perverse, one can see incompetent or entirely nonproductive people continuously employed for the sole reason that they were a constant source of adoration and an instrument of gratification for their *dehumanizing manager*. Workers may inform the manager that the individual is incompetent and unable to perform the job and the burden of work thereby increases for the others. But when the employee is a source of gratification for the *dehumanizing manager*, then any evidence that demonstrates incompetence will be rejected and ignored. At that point, the other workers may become disillusioned as they see that their knowledge, experience, and advice have been treated with disrespect. The *dehumanizing manager* manipulates the situation, presents events in his favor, declares that the employee in question performed well on some (trivial) piece of work (which probably no one else ever saw), and uses it to justify retaining the individual. Workers thus may subconsciously resign themselves to roles as the subordinated drones of their *dehumanizing manager* without even realizing what they have accepted. They show complete resignation with passive remarks, such as, "What can I do?"

It is dangerous to allow such people to arrive at or remain in managerial positions, or any position of authority. The resultant demoralization of personnel, decrease in productivity, and the onset of chronic psychological distress (and subsequent illness) among workers starts very early among *dehumanizing managers'* subordinates. The first symptoms of psychological distress in the victimized subordinates of *dehumanizing managers*, depending on individual susceptibility, include heart palpitations, feelings of oppression, anxiety, breathlessness, fatigue, insomnia, nervousness, irritability, headaches, digestive ailments, and stomach pain. Susceptibility to stress-induced symptoms can be acquired gradually when an individual is faced with chronic abuse. The victims do not see an escape route because they do not comprehend the on-going process. Nothing makes sense when contradictions abound and evidence is denied.

Chronic psychological distress ultimately attacks the neuroglandular function. With exposure to the chronic distress of working under a *dehumanizing manager,* the physical organism's resistance (and immune system) wears down and chronic and generalized anxiety, panic attacks, fear, fainting spells, and depression may appear. They can be difficult to control, leaving the worker in a permanent state of tension and hyper-alertness. Such managers give ulcers to their subordinates. (What has come to be called the "executive stress syndrome" is, in reality, a myth.) "In general, executives of all species are more likely to be giving ulcers than to be getting them" [37].

It also is well demonstrated that people in higher levels of management usually have lower rates of coronary heart disease than do people working at lower levels [4, p. 256; 37]. Functional and organic disorders can occur because of the attack of chronic stress on the neurohormonal system [14, p. 156]. In time, the worker's system weakens. They lose their capacity to resist, growing psychologically and physically exhausted, eroded as a human being, their spirit and identity destroyed. At this stage, adaptive techniques no longer work and long-term or acute illnesses may set in. They include hypertension, attacks of amnesia, rashes, cardiovascular disease, cancer and suicide, all known effects of the chronic distress induced by work under a *dehumanizing manager*. The effects of such experiences are not short-lived; indeed, they become part of one's psyche.

Victims say that ten or twenty years later they still feel distressed when they see images of their *dehumanizing manager*. The important long-term consequences of abuse victims in an organizational structure often are perceived only when, after a long absence from the job, it is suggested they return to work. This re-triggers the symptoms—anguish, insomnia, dire thoughts—and the person enters a vicious cycle of relapse: leave from work, back to work, relapse, and so forth, which can result in permanent rejection [14, p. 165].

Dehumanizing managers are a frightening act to watch in positions of authority, where they respect no one and create tremendous on-going daily distress and chaos for those working around them. In managerial or other positions of decision-making authority over others, *dehumanizing managers* bring dysfunction and tremendous costs to their employer where they use the system for their own benefit.

How do they get into managerial positions? *Dehumanizing managers* work hard to climb up hierarchical work ladders. They need to use and control more and more people. A *dehumanizer* uses seduction to get into a management position. Once there, usually no one is stupid enough to pick a fight with him. That is why they are so often found in managerial positions and left in those positions, as human wrecking crews.

Dehumanizing managers spend most of their time striking out at everyone around them at work. One would think that the person quite simply needs to be hit back from time to time by people who will show that "this bully or abuser cannot cross the line with me." Unfortunately, that approach does not work because *dehumanizers* are immune to talk of limits. The only way to set limits with a *dehumanizing manager* is by scaring him. And the only thing that scares such a person is any damage to the self-image he needs to project.

DEALING WITH DESTRUCTIVE MANAGERS: ACTION NEEDED

If one wishes to preserve and maintain the relationship with a *dehumanizing manager*, do not criticize or disagree with him. Provide him with ample and recurrent admiration, attention, affirmation, and applause. Never give advice. Stay away until summoned. Do not have a full-fledged existence or needs of your own. This is an impossible order for any working person to fill and one that does not belong in the workplace. People are paid to do a job, not spend their work days catering to the emotional needs of destructive managers. There are only two ways to end this terrible and worsening catch-22: either by radically changing the dominant model of management practices, putting people first and not profit, or by raising awareness in the workplace so that working people at all levels know what to look out for and have employer commitment for action when needed. Employer commitment to action must focus on prevention, training, raising awareness, and removing *dehumanizing managers* from managerial functions where they have any decision-making authority over subordinates.

What about close monitoring of the actions of such people? This approach is time-intensive and costly and requires trained personnel to closely monitor all decisions and actions taken by the manager in question. How effective monitoring can be is highly questionable. Would subordinates wish to work under a person identified as needing close scrutiny? It would reflect a general lack of confidence in the person's actions and decision-making capabilities. Furthermore, given that *dehumanizing managers* can be predicted to co-opt the system and not respect limits, using threats (such as removal from the job, use of some company sanction, use of legal action, and firing them) to scare them into towing the line does not appear to be realistic.

Workers must be able to establish limits with a *dehumanizing manager*, to tell the manager when he has reached the line that may not be crossed. But without the fear of action against the manager when he will not stop, this approach is useless. At the first sign of dehumanizing treatment from a manager, it is critical to note any provocation or aggression. One must collect all traces and indications of abuse, write down a description of the offenses, and make copies of evidence that supports the defense.

The following is an example of how one female airport check-in worker in our study "nipped it in the bud" with a *dehumanizing manager* while performing ground staff functions.

> At one point I worked for a man who was a screamer, a really destructive guy. Everyone knew it. As a manager, he often would close his door and chew someone out. The entire floor could hear him. I was one of his direct subordinates. One particular day, when he was extremely worked up, it was my turn. It was early evening, and many of my co-workers had already left. My office was the last office in a small hallway: dead end, a trap. My boss came into my office and was very angry about something. The veins were bulging out of his forehead. I thought to myself, "uh oh." He was standing in my doorway so I could not get out of my office. I was standing behind my desk. The situation deteriorated quickly. I told him we should talk about it tomorrow; I asked him to calm down and stop yelling. He moved to close the door. That scared me. I asked him to leave the door open. His yelling got louder. I prayed someone would come down the hallway. He kept shouting. His face got redder. He was physically in my office, approaching me. I got really, really scared. Somehow I mustered the courage to bolt past him while he continued raging. Once out of my office I said, "You are out of control. I am not going to stay here. You are scaring me. I am leaving."
>
> I left, shaking and shaken. After that I never permitted myself to be alone in a room with him. After that, if he was the only other manager on the floor in the evening, I would not work. If my co-workers left, I left. If my boss requested a meeting, the door had to be open. I made sure our meetings were not subject to his control—open doors, other people around, and I sent a written statement to the Human Resources Department. He was a smart guy, but an extremely abusive manager. Fortunately for me, he was not my boss for very long after that—restructuring took care of that. (Check-in worker)

This young woman had the tools to recognize the pattern of behavior and was prepared to take action against her boss to put an end to his unacceptable, disrespectful, and destructive treatment toward her. During an interview, I discussed with this worker her views on how she obtained the strength, resolve, and knowledge to identify unacceptable behavior from her manager and to act against it. Her ability to "nip it in the bud" appeared to come partly from self-determination and a natural inner strength, partly from her refusal to tolerate being shouted at, put down, or dehumanized in any way, and partly from her own resilience. The worker in this example understood that she could not work as a subordinate to her destructive manager and that a job change was required.

TALLY THE COSTS

If companies started to tally the costs of the damage done by such people when they are in managerial positions, and the costs related to workers' distress, MSDs, cardiovascular disease, cancer, medical treatments for related health problems, as well as lost work time, reduced productivity, reduced morale, reduced profit, then they might understand the importance of changing their management practices and examine more closely the kinds of people allowed into managerial positions [38].

If work-related distress, strain and their negative effects are to be prevented or at least diminished, then one of the key means is through empowerment-oriented training and raising awareness at all levels. Teaching tools that will help workers steer clear of, or deal better with, destructive characters (particularly those in superior positions and especially those fitting the description of a *dehumanizing manager*) can help raise the proverbial bar to improve work conditions and reduce associated work-induced stress.

There are lessons to be learned from management successes in identifying and removing *dehumanizing managers*. "D" (to maintain her anonymity) owns her own management consulting firm. CEOs of top companies compete for her services. The CEOs call her "an iron fist in a velvet glove." She has excellent people management and organizational skills and a highly developed sense of intuition and people radar, both of which she uses with confidence. (Two qualities managers are usually afraid to trust are intuition and faith [27, p. 121].) "D" has consulted for many companies that suffered from dysfunctional and destructive management and which therefore experienced low productivity and insufficient or no profit.

Most companies that bring in "D" at crisis stages face crises precisely because of *dehumanizing managers*. "D" is brought in to "save" these companies. Sometimes she is able to save the company, sometimes not. Poor management practices and destructive managers are arguably the main causes of a sinking workplace, company, or organization. Top management usually does not want to see what is happening until it is too late.

"D" described to me one company in particular that she tried to "save," a company that was sinking because of a *dehumanizing manager*.

> The company had opened a plant in another country, where labor was cheaper. They sent a crew of their senior managers (all men) to set up and run the new foreign plant. The local work force was highly skilled, and much more educated than the work force back home. But these imported "managers" went into full force exercising their authority and establishing strict hierarchy. The general manager hired his brother-in-law as the production manager, and his son-in-law to work in a key position in the sales department. This general manager answered to no one, and did whatever he wanted. A classic dehumanizing manager, he co-opted the system for his

own benefit, using everyone as his pawns. The new locally hired workers were in a position of extreme disadvantage—how could they resist his seductions or refuse to carry out his follies-de-grandeur when he had hired them, in a company that appeared (at first) to be better than their local companies. And being far away in another country, in the head position, there was no one to see or control what the general manager was doing. And on top of it, he had the complete trust of the CEO.

Of course, workers had no involvement in how any of the work processes were designed or implemented, or in the company's decision-making—a particularly foolish and costly corporate transgression for a company on culturally foreign territory. Logic would dictate "when in Rome, ask a Roman," but not so with a dehumanizer for a general manager. He knew everything, he could never be wrong, for him no one else seemed to exist, no one else's knowledge or experience had any value. Not surprisingly, in less than one year the company was practically not functioning, and certainly was not making any money. The CEO hired me to try to save the factory.

On my first visit to the factory, I interviewed every person in the factory—workers and management. At the end of my first visit, I told the CEO that the first action he had to take was to fire the general manager, who I explained was an extremely destructive individual. I explained that he was doing irreparable damage to the workers, to other managers, and to the company, including to its reputation in that country. Naturally, the CEO refused, resisted, screamed, cried, complained, and made all kinds of loud noises. After all, this was his general manager, the man had been with the company for 25 years, and on top of it, the production manager was the general manager's brother-in-law, so firing the general manager would create a mess. "How can I possibly fire him?" the CEO asked me in desperation. But after a couple of months of working closely with the CEO, reviewing with him again and again the damage that this dehumanizing general manager was doing and was going to do, I succeeded in getting the CEO to fire his destructive general manager.

Immediately after his removal, things started improving. One of the most noticeable immediate changes was that both workers and managers stopped living in fear, no longer subjected to constant emotional upheaval or submerged in the destructive general manager's follies-de-grandeur. ("D", management consultant)

The next most immediate and important change "D" introduced was full participation of the entire work force in all company decision-making. Without worker participation, companies and organizations may make profit, but will never achieve their full potential as long as workers are treated as drones and incompetents who are incapable of making decisions for themselves, as people who know nothing of value to management, as long as management is allowed to operate with total authoritative, autocratic control. The full creative initiative and potential of workers cannot be achieved as long as management obliges workers to work by fixed and narrowly defined job descriptions, rather than fully expecting that most workers can and want to go far beyond those limits.

With further changes, that particular company quickly became strong, generating higher profits. Today, it continues to be a successful operation, providing employment for many local people who previously had been unable to find jobs.

Perhaps if everyone refused to work with *dehumanizing managers* it would send a strong message about removing them from positions where they can do harm to others and to the enterprise. But it requires a critical mass of people becoming aware of and skilled at recognizing such personalities early on and prepared to take action. Unfortunately, Human Resource Departments generally are not filled with people trained to recognize them, nor are most workers, other managers, or CEOs. Where human resource professionals do have the skills to recognize *dehumanizers*, all too often they take no action because they see their role as carrying out the word of management rather than acting as a neutral body to serve the work force and management equally. "We had an awful manager but no one dared complain to [H]uman [R]esources. We couldn't stand him. It was harassment all the time. I think more people called in sick then." (Check-in worker)

> Obviously one does not drop dead on the spot as a result of these aggressions, but one does lose a part of oneself. One gets home every night worn out, humiliated, and damaged. It's difficult to recover. . . . In critical moments we tend to emphasize already existing factors: a rigid company will become more rigid, a depressed employee will become more depressed, an aggressor more aggressive, and so on. We become caricatures of what we already are. A crisis situation can definitely push an individual to give it his or her all to find solutions, but emotional abuse anesthetizes the victim and only aggravates his or her characteristics [14, p. 53].

Workers have their intuition; very often and early, they can detect when a manager is destructive. Those inner voices must be listened to if workers are to take steps to protect themselves before they become trapped. Not everyone has come up against this kind of destructive person and may not be familiar with the signs. And no one likes to think of himself as having been used as someone else's tool. Because *dehumanizers* treat no one in the workplace with respect and see everyone as mere tools, their behavior often constitutes psychological harassment. It is at this stage that the health-related costs start to mount, both for individuals and the organization. Though the *dehumanizing manager* does not see that his behavior was wrong does not mean he should not be held responsible. Indeed, he should be held accountable. Whether or not such individuals are conscious of their own brutality, their behavior is nonetheless unacceptable.

When a manager says that a worker does not respect the manager's authority or complains that the worker is a "difficult person," it should not be assumed automatically that the manager is correct. That may indeed be the case, but it also may be true that the manager is abusing his authority, making such statements and judgements as a manipulation and a means of (further) *dehumanizing* the worker. When a manager makes negative comments about one or more subordinates, it should sound an alarm signaling an urgent need to scrutinize both the manager (who may be a *dehumanizer*), and possibly the worker(s). Managers who complain about their subordinates and try to remove them are almost certainly unable to manage or to encourage workers to achieve good performance levels. Basically, the burden should not be placed on the worker with the manager's word carrying all or most of the weight because a *dehumanizing manager* will manipulate to convince others that what they say is true. It happens too often, with the worker somewhere in the background, victimized by the situation, unable to manipulate as the manipulator does, appearing weak and unable to stand up to the arguments of the manager.

It is curious that companies still are largely unaware of the direct impacts of such individuals on productivity and profitability. When looking at ways to increase productivity, pressure is usually put on workers to work harder and faster to produce more. Yet in reality, the problem often is with the manager, who has had a de-motivating effect on the workers or who creates chaos that reduces productivity over time. If managers are not scrutinized, then this cause and effect can easily be overlooked and the problem never solved. Many companies and organizations suffer from low productivity and low profitability for this reason, or they simply do not function above the minimum level. Workers feel motivated when they feel appreciated, encouraged, and valued. If managers do not provide this emotional nourishment, then workers cannot be expected to perform their best. A *dehumanizing manager* is unable to provide a sense of appreciation or encouragement to his subordinates.

DEHUMANIZING MANAGERS ARE PROLIFERATING

People with this destructive nature are proliferating in our society for reasons that are structural, cultural, and personal. The result is that there are more of them in workplaces, especially in managerial positions.

One reason that *dehumanizing managers* are proliferating in workplaces is that the work world today has become increasingly merciless, operating by a sort of natural selection that allows *dehumanizers* to be placed in strategic positions. The structural and cultural hegemony dominating today's society and workplaces is underpinned by a throbbing drive to accumulate unlimited capital at any cost. Accumulating capital is used to justify the push for ever-greater productivity and work speed, the intensification of work processes, fiercer competition, treating workers as managers' tools, and getting more and more done with fewer and fewer people. Because *dehumanizing managers* are skilled at competition, especially through manipulation and the use of other skills lacking in ethical value, the dominant structure and culture serves to perpetuate their proliferation and ascension to managerial positions and to condone their *dehumanizing* behavior. Organizations do not seem to realize that the very people they place as managers are the cause of much ill health in the work force. The "hegemonic" behavior of such managers contradicts any semblance of dignified work or respectful treatment of workers as human beings.

Because *dehumanizers* are cold calculators, they take advantage of rational elements in a workplace without allowing themselves to be touched by human emotions. Generally speaking, such people are adept at making their way up in an enterprise, organization, or administration because they play dirty and are skilled manipulators and seducers. Someone who plays dirty is always likely to win, unless stopped. Organizations, as any place where there are power plays, have a tendency to attract *dehumanizers* and to give them a large playing field. The danger with such individuals lies not only in their personal behavior but also in their power of seduction because they know how to sweep entire groups along into their perversions [26, p. 349].

TAKING ACTION

The labor market flexibility dominating today's global markets means that far fewer working people spend their entire lives working for only one employer. With increasing labor market and employment insecurity, moving from job to job is the norm for most workers. The dominant model of hierarchically structured workplaces based on competition and the fact that there are more workplaces today than in previous periods of human history means that there are more people in managerial positions than ever before. It is not an accident. More layers of managers have been introduced in workplaces world wide as part of the structural and cultural hegemony to exercise more control over working people. Even with trends in some countries and in some sectors of the economy to reduce layers of management by creating more "flat structures," control over working people has not lessened concomitantly. To the contrary, control has heightened through work intensification, longer hours, reduced employment security, increased income insecurity, and fewer collective agreements negotiated at the workplace level due to reduced collective voice (that is, diminished membership in trade unions).

People face an ever-increasing risk of working for a *dehumanizing manager* at some point during their work lives. Learning to cope is essential to prevent workers from falling ill and to mitigate the associated costs.

This section discusses what to do when faced with a *dehumanizing manager*, while the next section addresses strategies for prevention.

There are only two possible courses of action: remove the manager, or remove the worker (preferably through collective action). When a subordinate is faced with a *dehumanizing manager*, if the individual does not have support for collective action, then the most important action is to run the other way. All too often, job change for the subordinate of a *dehumanizing manager* is seen as the solution, including by the worker. However, obliging the worker to move away from the destructive manager is a morally and ethically indefensible solution. It is the cowardly approach by top management. An easy recommended solution is silent acceptance of the problem and to look for another job—except that the next victim is already on the list. Removing oneself from the path of destruction is essential. However, there are consequences of running the other way. If we all run from *dehumanizing managers*, then we allow them to take over. It is they who must be removed and prevented from working in managerial positions, but most of the time that is not what happens.

Let us examine a few key reasons why removing the worker is a non-solution: 1) it is the manager, not the worker, who has created intolerable work conditions; 2) the worker may want to keep her job but may not see any option other than moving away from the *dehumanizing manager*; 3) moving the worker and not the manager implicitly condones the behavior of the manager, which is an invitation for a repeat performance; 4) moving one worker leaves the other subordinates of that particular *dehumanizing manager* as sitting ducks; 5) moving one worker and leaving the other subordinates to still have to deal with the destructive manager will eventually result in serious de-motivation of the work group and guarantee productivity loss.

Dehumanizing managers generally are not relieved of their positions for several reasons. What does leaving *dehumanizing managers* in their positions say about the top management and CEOs in companies and organizations where these people work? The fact that top management and CEOs do not remove such destructive people from positions where they can do irreparable damage usually indicates any or all of the following:

1. Both individual and collective cowardice on the part of other managers;
2. That those in top management, including the CEO, are not *dehumanizers,* and therefore do not know what to do when they find themselves facing a *dehumanizing manager* [31, p. 258; 39].
3. That those above the level of the *dehumanizing manager*—including CEOs—are also *dehumanizers*, pure opportunists, concerned only with themselves, and devoid of interest in what happens to members of the work force, even when it reflects on the organization. In such a case, that CEO or top manager will weigh only the consequences to himself of taking action to remove such a manager. This scenario is a vicious circle between two *dehumanizers*. If the CEO or top manager thinks he will have a hard time when he removes the *dehumanizer* (a law suit, for example), then he probably will take no action, thereby sacrificing an entire cohort of workers by inaction. All workers are potential sacrificial lambs for *dehumanizers* at any level;

4. The "old boys' network" is in action, a key mechanism used to establish and perpetuate the patriarchal allegiance among men at junior and senior levels of management. The "old boys' network" is used quite effectively to keep men in their managerial positions no matter how bad or destructive they are [40];

5. Top management does not wish to "lose face" because they probably are the ones who appointed the manager. Removing him would reveal a certain lack of competency on the part of top management, which can lead to a loss of confidence in the company or organization by people both inside and outside. (There are also ways to present "removals" to those inside and outside the company or organization that actually leads to greater confidence in the organization.)

It is irresponsible inaction that managers and human resource professionals are not pro-actively trained to recognize such people before they ever get into decision-making positions, in order to prevent damage before it occurs.

The preferred course is collective action. An individual alone should not attempt to fight or defend against a *dehumanizing manager* because the *dehumanizer* can be counted on to lie, play dirty, and be vindictive. Unless you are like them, the fight will always be unfair, and you will lose. Only collective action can create change [41]. Collective action should work toward: 1) ensuring that companies and organizations provide training for awareness-raising and skills development at various levels of any workplace; 2) ensuring that destructive managers be removed from positions of authority; 3) ensuring that top management commits to and fully supports points 1 and 2; 4) ensuring that personality assessments are used to prevent destructive people from getting into managerial positions; 5) ensuring that union management receive education and the necessary tools to intervene in dysfunctional or organizational situations before they reach the dismissal stage. Employees must react quickly and together against any manager who treats workers in a *dehumanizing* manner. They should request intervention from the union, the labor inspectorate, and the Director of Human Resources. Occupational physicians in the enterprise have a critical role to play as well to help protect workers who are victimized [26, p. 354].

Additionally, organizational policies addressing sexual harassment should not be separated from policies addressing psychological harassment. Harassment is a form of abuse, of violence, and it must be recognized that sexual harassment always involves some degree of psychological harassment as well. Establishing separate policies for these two forms of harassment, or having a workplace policy only about sexual harassment actually works against victims, and ultimately does not help employers to address the problems. Policies against harassment are necessary, however they should encompass the larger picture of harassment and its implications of abuse. Sexual harassment can be said to be one expression of psychological harassment, or psychological violence.

In the work world generally managers, CEOs and Human Resource professionals do not take seriously enough the difficulties in work relations, unless they become an issue that affects the image of the organization. For the most part, managers prefer to avoid dealing with relational difficulties between people. The rage or frustration that accumulates, fueled by small conflicts between people, causes enormous silent suffering and can result in many workers falling ill. A large study in France revealed that 19% of workers who reported suffering under *dehumanizing managers* sought help from their organization's Human Resource Department, but only 1% received any help. Similarly, 37% of the workers who suffered from *dehumanizing managers* sought help in the hierarchy where they worked, but only 5% received any help at all [26, pp. 359-360, 368]. Even when they are aware of the reality of the problem with managers, Human Resource Department professionals often vacillate between denial, seeing the situation as banal, and being perplexed. In general, their position is one of ambiguity. In principle, since Human Resource Departments are meant to be the intermediary between workers and management, they are the best suited to put *dehumanizing managers* in their place. However, in reality, the departments typically merely reflect the instructions of top management in a neutral manner and without intervening. Although dealing with these situations may be difficult, nonetheless, it is unacceptable for a Human Resources Department to do nothing. People get hurt, sometimes with lifelong consequences, sometimes including death. At the minimum, there are the costs to consider if one reduces everything to the bottom line.

This points to the importance of the need for increased training and professionalism in the field of human resources. A Human Resources Department is the backbone of any company or organization and a mirror reflection of the qualities upheld and put forth by the CEO and top management. If top management does not emulate respect for its workforce, if it does not treat employees with dignity, then one should not expect the Human Resources Department to behave better in its dealings with employees. Management sets the tone, pulls the strings, and determines whether working life will be relatively harmonious, or carried out in discord. The impacts of Human Resource Department personnel have taken on new dimensions in today's cutthroat world of work. Where human resource personnel are not professionally trained, skilled and experienced, demonstrating the highest degree of integrity and decency, employees are likely to suffer and fall ill. The potential is so great for human resource personnel to create direct and collateral damage that I suggest there is a need for the field of human resources to liase with the field of public health. An individual's work-induced distress and ill health has negative impacts for families and communities. This becomes a population health issue, with cost implications at individual, enterprise, family and community levels.

In industrialized countries today, there may be less social conflict, but there is much more individual suffering in the workplace: distress, fatigue, anxiety, depression, and psychological abuse. Each individual suffers in her corner, unable to share her troubles with a cohesive group. Trade unions have great difficulty resolving the problems of individual suffering because they cannot be addressed only at a collective level. However, the problems related to *dehumanizing managers* and destructive management practices present a new organizing opportunity for trade unions. Unions must train themselves in this area while not denying the psychological elements of the problem. Unions must learn to not deny the individual, which happens when they see their role as focused exclusively on collective interest [26, p. 362]. Protecting individuals from the destructive effects of *dehumanizers* at work is, without question, working for the collective interest.

It is the role of the trade union to identify where individual managers are the source of distress and undue strain in their subordinates, which can lead to cases of psychological harassment with its associated damaging effects on workers. The union can discuss productivity objectives with top management through workplace committees. Notwithstanding actions or discussions with top management, the union must react immediately against mistreatment of subordinates by managers [26, p. 362]. It is incumbent on the union to question the *dehumanizing manager* and top management and to demand a change in managerial methods, which may include removing the *dehumanizing manager*. Once these measures have been followed, if the top management does not respond with actions to protect the workers and to change the situation, then public measures should be envisaged, such as a strike or calling in the mass media. Exposing individual *dehumanizing managers* in a public forum, including in the press, is extremely potent for individuals whose image represents everything to them. Equally potent is exposing publicly the organizations that allow such individuals to remain in positions where they harm subordinates [26, p. 363].

The reverse is equally effective: that is, to use the press, media, and organized groups to publicly promote companies and organizations that treat employees well and with respect, the places where workers generally are happy. Companies and organizations do not like to be conspicuously absent from lists of good employers, especially if their competitors or brethren organizations are lauded. Thus, instead of naming and shaming unhealthy workplaces, change can be catalyzed by naming and praising workplaces known to use good management practices.

STRATEGIES FOR PREVENTION

Managers, subordinates at all levels, and trade union members must receive training so that all personnel in a workplace learn the characteristics to look for to identify *dehumanizers* and the tools needed to set the proverbial bar sufficiently high. Unions should be fully involved in the design and implementation of the training at workplace level. Of no less importance, training should include a toolbox of skills necessary to deal with workplace *dehumanizers*, including putting a stop to the abuse of power and to discrimination. Legislation should be used to remind managers that abusive treatment constitutes violence in the workplace. When working with *dehumanizers*, health-related impacts should be compensated as for an occupational accident. And managers should be trained so that they are discouraged from being abusive, harassing, or in any way treating workers without respect. Training on the recognition of workplace abuse could be included in the meetings of workplace health and safety committees.

Tacit acceptance of a *dehumanizing manager's* behavior and of working people's resignation to this kind of ill treatment (accepting it once means you are stuck in the *dehumanizing manager's* web for good until you walk away) is reprehensible and unacceptable. It is indeed immoral and unethical in any workplace to remain indifferent to the problem of emotional abuse [33]. Ultimately, we are all responsible for letting this widespread problem poison so many relationships in the workplace.

Prevention is the most important means to "nip it in the bud." The tools necessary to "nip it in the bud" involve not only collective responsibility, but also recognizing that it is up to each one of us to question and not blindly follow the dominant system. It sometimes means saying "no," and learning to not accept passively anything and everything in the workplace. Learning to pinpoint what is good for oneself and to say "no" to that which runs in opposition to one's own moral values is primordial [26, p. 403]. One should bear in mind, however, that saying "no" can be especially difficult for women workers since women generally are socialized to not say "no" and to be accommodating, both at and outside work [40].

A critical measure of prevention is to educate people to deal correctly and respectfully with others at work. To introduce this kind of education, enterprises and organizations should inculcate in their work forces standards for decent and respectful behavior, define what the organization considers acceptable behavior, and what it considers unacceptable. It does not mean an enterprise should impose morality on its work force but it does require setting limits. An employer must ensure that each employee understands the potential consequences of his or her behavior toward others. Respectful treatment, however, has to be demonstrated first and foremost by the CEO and top management. Only then can it be taken seriously and "trickle down" to become the "workplace culture."

There is a great need to discuss and reinforce the fact that managers should earn the respect of subordinates; workers need not give their respect to superiors if it has not been earned. Respect within a hierarchy is not innate to humans, it is socialized—we are taught to respect superiors without questioning. Managers must earn what they feel is their due. Collective pressure based on workers' demands to work with managers who inspire workers' respect would result in the natural attrition of destructive managers.

It is a challenge for working people to identify emotional abuse or unacceptable, disrespectful treatment of any degree in the workplace, to break its destructive spell, and to learn to work in relationships based on respect and decency rather than power and cruelty. Decent, dignified work cannot co-exist with relations based on power and cruelty.

To recognize *dehumanizing managers* is to guard against them. It is a mistake to see relational conflicts in the workplace as something to be avoided at all cost. Conflicts present an opportunity for change and to gain awareness of others. Abusive or destructive treatment in the workplace blocks all possibility for positive change. Unacceptable treatment creates both inertia and a downward spiral in the mental and physical health of those who are treated without respect and dignity.

If working people expect decent work conditions and to be treated with dignity and respect at work, then it follows that *dehumanizing managers* must be identified and removed from positions where they can inflict harm on workers at any level. The act of removing them will make other *dehumanizing managers* think several times before inflicting cruelty on other unsuspecting subordinates.

There is an urgent need to discuss openly and widely the fact that all forms of harassment in the workplace constitute psychological violence, so that when making policies to protect workers from violence at work, these pervasive forms are recognized and included. It is important to make these issues better known so that workers can walk away from the unacceptable.

No one can ignore the difficulties of work environment or job change when you feel you cannot or will not take any more. The body of evidence shows that when there is a separation of subordinate(s) from a *dehumanizing manager* (unless the *dehumanizing manager* is removed from his position of authority over others) it is usually the subordinate who leaves. And during the separation process, the *dehumanizing manager* always considers himself the wronged party and often becomes litigious.

> It's not unusual for victims to be sued in an organizational framework, because they are always considered guilty. Either way, the abuser claims injury, whereas the victim is the one who loses everything [14, pp. 162-163].

These situations should be avoided by "nipping in the bud" these problems before workers get sick and organizations become enmeshed in drawn-out lawsuits. A multitude of examples remind us that collective voice can catalyze positive change [42]. In fact, often it is the only available mechanism. Training, awareness-raising, and simple discussions in the workplace can help working people at all levels to know that they must protect themselves and should demand their rights. Collective agreements, negotiated jointly by union and management, are needed to support workers' rights.

> [Dehumanizing managers have] such power to do harm that it is difficult to contain. If individuals and organizations don't find ways to instill civility and respect for other human beings, it will become necessary to legislate emotional abuse based on the model of sexual abuse [14, p. 179].

The minimum of responsible action is for employers to conduct a thorough psychological screening [43] of potential managers, given the responsibility such people will have and the potential for them to do harm. This simple preventive measure can be applied with a high degree of effectiveness and reliability using personality assessments that identify the behavioral dimensions of destructive people.

The high risk for people in positions of authority to lose their powers of perception is a strong argument for using tools such as personality assessments to help screen applicants, particularly to identify the behavioral dimensions of *dehumanizers*. Assessments that contain the behavioral dimensions of destructive people can: 1) identify people who should not sit in managerial positions (preventive action to prevent sickness in workers and costs to employers and workers), and 2) identify those who got into managerial positions, but who should be removed (curative action to prevent or help treat sickness in workers that was induced by destructive managers). Identifying destructive people preventively (or curatively) is healthy, sensible, and responsible employer action for any enterprise.

The application of personality assessments can be quite simple but requires involving a person trained in using them. There are five behavioral dimensions that are sufficient to identify people who as managers will create an unhealthy environment for others and for the enterprise or organization [44].

Behavioral Dimensions of Dehumanizers in the Workplace

1. Excessively high degree of self-confidence, no empathy, does not care about others.
2. Extrovert, someone who shows off.
3. Not modest, talks about his successes most of the time, or claims others' efforts as his efforts.
4. Not sensitive, does not consider other people's criticism, always finds excuses or blames others for his own mistakes.
5. Not democratic, does not care about the ideas of others, does not trust other people's thinking process, thinks his own thinking process and behavior is the best, always insists on doing everything his own way.

The Manager-Worker Dichotomy:
An Unproductive Relic of the Past

In addition to using personality assessments, workers should be involved in decision-making. Workers usually have to be evaluated each year by their supervisors. It is both logical and necessary for managers to be regularly evaluated—twice a year, for example—by all workers they supervise, to ensure managers' correct performance on the job. Some companies use this practice with great success (Semco, Brazil, is an inspiring case of a democratic workplace where the manager-worker dichotomy is seen as an unproductive relic of the past), but it is far from being the norm. In most workplaces, if worker evaluation of managers is proposed, the suggestion is likely to be rejected immediately by the managers, revealing their fear of loss of control. But the cycle of autocratic decision-making has to stop somewhere. If evaluations come from the top down, they also must come from the bottom up. It re-emphasizes the importance of discussing and making less distinct the manager-worker dichotomy.

> In these days of the new world order, almost everyone believes people have a right to vote for those who lead them, at least in the public sector. But democracy has yet to penetrate the workplace. Dictators and despots are alive and well in offices and factories all over the world [27, p. 170].

Workers evaluating managers is a healthy practice in any workplace and benefits everyone in the organization. This simple practice brings major cost savings to companies in the long run because ineffectual managers are revealed after only one or two evaluations. Without this kind of evaluation, ineffectual managers often coast along using other mechanisms to hide their limitations. A major additional benefit of worker evaluation of managers is that it is a simple and obvious mechanism for recognizing *dehumanizing managers*. It also can save the enterprise enormous sums of money by "nipping in the bud" within six to twelve months those destructive managers who are nearly guaranteed to make workers sick; who cause workers to live in fear, thereby rendering them unable to produce or work to their potential; who will co-opt the enterprise, thereby misusing its resources; or who will single-handedly cause the enterprise to lose profit.

Twice yearly worker evaluations of their managers is a sensible and obvious mechanism:

1. It is a clear means of involving workers in decision-making and a good place to begin worker participation where it does not exist;
2. It is a powerful cost-saving device for any employer;
3. It is a way to eliminate the potential need for all kinds of legal mechanisms and their associated costs. Such mechanisms often are put in place because of the destructive practices of some managers leading to harassment cases by workers or due to legal cases stemming from manager-induced distress;

4. It is an efficient mechanism for reducing, even eliminating, work-related distress and associated ill health effects, saving employers, workers, and their families from tremendous costs and suffering;
5. It is a way to keep improving the quality of management in the enterprise or organization, helping managers to learn and grow;
6. It leads to much better relations and communication in the workplace;
7. It is a way to keep improving the quality of workers in the enterprise or organization, helping workers to learn and grow;
8. It simply makes good sense and is an ethical practice.

Worker evaluations also could be linked to workers surveying their workplace for safe work environment practices.

In addition to workers evaluating their supervisors or managers, workers also should have full voice and decision-making authority over who is recruited into their work group—including a new supervisor or manager—because the workers are the people who will have to work with the new person. This very obvious and inclusive practice unfortunately is used rarely in most workplaces, but successful companies have learned that it is necessary for a company that includes workers in decision-making.

Workers know their jobs better than anyone else [36]. By interviewing potential recruits, they can understand better than anyone whether a recruit is qualified for the job and whether the person will contribute to the work group or has merely talked their way past management. Because workers are usually excluded from the decision-making process in recruiting, it happens all too often that management brings an inappropriate person into a work group. Upon close examination of what happens in work groups, one can see that uneven workload is often the direct result of other workers having to carry out the work, or any part thereof, of someone who is not suitable for their job. In sum, more competent workers end up carrying the burden for someone who does not pull his weight. The stress and resentment this creates over time for those performing the job of another has consequences for the individual affected, the work group, and the company.

The same principle applies to introducing a new supervisor or manager into an existing work group. The workers in the group should interview all candidates and make their own selection. This approach is a simple, very effective tool for improving the quality of managers. Those who do not make the grade in the eyes of the workers will not last long—again, natural attrition.

When workers are involved in the recruitment of new personnel or managers, the decision of the majority of the group should never be overridden, i.e., top management should not go against the decision of the majority, but management also should have a voice (a vote) in the recruitment of personnel, including managers.

Dehumanizers May Feel Pleasure in Giving Pain:
Reason to Involve Workers in Decision-Making

The full involvement of workers in recruiting processes is a simple step to help screen out *dehumanizers* before they get into managerial positions. Workers may need basic training to raise their awareness about the striking characteristics of *dehumanizers* and to know what to look for. They should, at minimum, receive training on the key behavioral dimensions of *dehumanizers*.

Management training courses do not help *dehumanizers* [45]. Companies may not want to lose the talents and skills of such individuals, therefore another job should be found for them bearing no responsibilities for supervising people. (But, when an enterprise has acted so, it should be made known that the person is removed from managerial functions because they are destructive and abusive to subordinate staff).

It is entirely possible that *dehumanizers* feel the sensation of pleasure when they give pain to other people. Humans seek to repeat behaviors that give them feelings of pleasure, which may explain the repeated patterns of destructive behavior in *dehumanizers* [46]. This dimension has been described in neuroscience literature but more research is needed.

If we do not try to understand something about the psychology of what drives people to act as they do in their jobs, particularly those in decision-making positions, then we cannot know how best to tackle direct stressors and resultant problems. Attempting to understand something about the truly destructive nature of particular types of people in the workplace is important to comprehending how they use or abuse various management practices put at their disposal. Because their actions have direct impacts on working people, employees must understand not only the practices used by their organization's management, but also how some individuals abuse those practices.

Management practices that exclude workers from decision-making leave workers feeling humiliated, degraded, treated without respect, their ego needs unfulfilled and unsatisfied. Chronic stress mounts, with pernicious, sometimes deadly effects. One cannot say that a workplace where *dehumanizers* are found in managerial positions is a safe and healthful workplace. One can say that such a workplace is an unhealthy place to work. In workplaces with a strong culture of promoting and protecting the well-being of workers, *dehumanizers* will not be found in decision-making positions with authority over others.

We cannot divorce the widespread impacts of destructive managers from today's dominant management practices. It is systems of work, i.e., management practices, which allow destructive individuals to fill management positions, implement *dehumanizing* hierarchy, and exclude workers' voices from decision-making. Today's socio-economic system cultivates the flourishing of such practices and destructive individuals. Several decades ago, the cornerstone of building a career was competency, not manipulation and seduction. Today, seduction and manipulation are all too often the major tools used to climb a corporate ladder into managerial positions. If nothing else, the proliferation of workplace manipulators indicates the ascendancy of the individualism that dominates Western societies. Manipulating seems to be the status quo for career advancement in the United States and Western Europe, highlighted by the gradual renouncing of behavioral constraints and prohibitions under the guise of tolerance.

A rallying cry in the 1970s advocated questioning authority. An inquiring mind should question everything in the workplace (as long as the questions are valid and you are ready to prove that what you say is right). Should things be done a certain way just because "that's the way it was always done"? That is the usual answer, but perhaps that is a lazy person's answer. Thinking of the brain like a muscle, it too can atrophy if not exercised. Asking questions and critically analyzing is good brain exercise and should be encouraged in the workplace.

ENDNOTES

1. J. King, Bad Boss, Bad Morale, *Computerworld, 38*:45, p. 54, 2004.
2. B. E. Ashforth, Petty Tyranny in Organizations: A Preliminary Examination of Antecedents and Consequences, *Revue Canadienne des Sciences de l'Administration, 14*, pp. 126-141, 1997.
3. M. K. Duffy, D. C. Ganster, and J. D. Shaw, Positive Affectivity and Negative Outcomes: The Role of Tenure and Job Satisfaction, *Journal of Applied Psychology, 83*, pp. 950-959, 1998.
4. S. V. Kasl and S. Cobb, Psychological and Social Stresses in the Workplace, in *Occupational Health: Recognizing and Preventing Work-Related Disease*, B. S. Levy and D. W. Wegman (eds.), Little Brown and Company, Boston/Toronto, 1983.
5. H. Knudsen and S. Tinney, Workplace Abuse, Job Stress, and Alcohol Outcomes: An Exploratory Study, in *Southern Sociological Society*, 2002.
6. C. Traweger, J. F. Kinzl, B. Traweger-Ravanelli, and M. Fiala, Psychosocial Factors at the Workplace—Do They Affect Substance Use? Evidence from the Tyrolean Workplace Study, *Pharmacoepidemiology and Drug Safety, 6*, pp. 399-403, 2004.
7. R. Folger and R. A. Baron, Violence and Hostility at Work: A Model of Reactions to Perceived Injustice, in *Violence on the Job: Identifying Risks and Developing Solutions*, G. VandenBos and E. Q. Bulatao (eds), American Psychological Association, Washington, D.C., pp. 51-85, 1996.
8. S. Moore, L. Grunberg, and E. Greenberg, A Longitudinal Exploration of Alcohol Use and Problems Comparing Managerial and Nonmanagerial Men and Women, *Addictive Behaviors, 28*:4, pp. 687-703, June 2003; R. Bhagat, D. L. Ford Jr., M. O'Driscoll, L. Frey, E. Babakus, and M. Mahanyele, Do South African Managers Cope Differently from American Managers? A Cross-Cultural Investigation, *International Journal of Intercultural Relations, 25*:3, pp. 301-313, May 2001;

L. Tetrick, K. Slack, N. Da Silva, and R. Sinclair, A Comparison of the Stress-Strain Process for Business Owners and Non-Owners: Differences in Job Demands, Emotional Exhaustion, Satisfaction, and Social Support, *Journal of Occupational Health Psychology, 5*:4, pp. 464-476, October 2000; J. Portello and B. Long, Appraisals and Coping with Workplace Interpersonal Stress: A Model for Women Managers, *Journal of Counseling Psychology, 48*:2, pp. 144-156, April 2001; N. Ghorbani, P. J., Watson, and R. Morris, Personality, Stress and Mental Health: Evidence of Relationships in a Sample of Iranian Managers, *Personality and Individual Differences, 28*:4, pp. 647-657, April 2000; M. Cavanaugh, W. Boswell, M. Roehling, and J. Boudreau, An Empirical Examination of Self-Reported Work Stress Among U.S. Managers, *Journal of Applied Psychology, 85*:1, pp. 65-74, February 2000; H. Frankel, M. K. FitzPatrick, S. Gaskell, W. Hoff, M. Rotondo, and C. Schwab, Strategies to Improve Compliance with Evidence-Based Clinical Management Guidelines, *Journal of the American College of Surgeons, 189*:6, pp. 533-538, December 1999; T. Witkowski and E. Thibodeau, Personal Bonding Processes in International Marketing Relationships, *Journal of Business Research, 46*:3, pp. 315-325, November 1999; U. Lundberg and M. Frankenhaeuser, Stress and Workload of Men and Women in High-Ranking Positions, *Journal of Occupational Health Psychology, 4*:2, pp. 142-151, April 1999; R. Peter and J. Siegrist, Chronic Work Stress, Sickness Absence and Hypertension in Middle Managers: General or Specific Sociological Explanations? *Social Science & Medicine, 45*:7, pp. 1111-1120, October 1997.

9. L. Keashly, V. Trout, and L. M. MacLean, Abusive Behavior in the Workplace: A Preliminary Investigation, *Violence & Victims, 9*, pp. 341-357, 1994.

10. J. Richman, Workplace Harassment, Symptomatic Distress, and Alcohol Use and Abuse among University Employees, *International Sociological Association*, 1998.

11. Abusive conduct includes any conduct through ". . . words, looks, gestures, or in writing, that infringes upon the personality, the dignity, or the physical or psychical integrity of a person; also behavior that endangers the employment of said person or degrades the climate of the workplace" [in 14, p. 52].

12. L. Keashly, Emotional Abuse in the Workplace: Conceptual and Empirical Issues, *Journal of Emotional Abuse, 1*:1, pp. 85-117, 1998.

13. S. Shahtahmasebi, Quality of Life: A Case Report of Bullying in the Workplace, *Scientific World Journal, 4*, pp. 118-123, 2004.

14. M.-F. Hirigoyen, *Stalking the Soul: Emotional Abuse and the Erosion of Identity*, Helen Marx Books, New York, 2000.

15. M. Kivimaki, M. Virtanen, M. Vartia, M. Elovainio, J. Vahtera, and L. Keltikangas-Jarvinen, Workplace Bullying and the Risk of Cardiovascular Disease and Depression, *Occupational and Environmental Medicine, 60*:10, pp. 779-783, 2003.

16. M. A. Vartia, Consequences of Workplace Bullying with Respect to the Well-Being of Its Targets and the Observers of Bullying, *Scandinavian Journal of Work, Environment and Health, 27*:1, pp. 63-69, 2001.

17. H. C. Ball, Management of Bullying and Bullying in Management, *Accident Emergency Nursing, 4*:3, p. 109, 1996.

18. J. Stebbing, S. Mandalia, S. Portsmouth, P. Leonard, J. Crane, M. Bower, H. Earl, and L. Quine, A Questionnaire Survey of Stress and Bullying in Doctors Undertaking Research, *Postgraduate Medical Journal, 80*:940, pp. 93-96, 2004.
19. D. Jackson, J. Clare, and J. Mannix, Who Would Want to be a Nurse? Violence in the Workplace—A Factor in Recruitment and Retention, *Journal of Nursing Management, 10*:1, pp. 13-20, 2002.
20. M. A. Lewis, Bullying in Nursing, *Nursing Standards, 15*:45, pp. 39-42, 2001.
21. A. Rutherford and C. Rissel, A Survey of Workplace Bullying in a Health Sector Organization, *Australian Health Review, 28*:1, pp. 65-72, 2004.

22. Additional reasons for the lack of discussion about the obvious association between destructive managers and the increasing toll of work-related stress, with its associated costs:

    • There is a lack of collective agreements in workplaces that provide support to workers, outlining, in black and white, the actions workers can pursue if they are victimized by destructive managers.

    • Researchers in the aforementioned fields (as in most others) typically do not plunge into the research found in other fields, such as psychology and psychiatry (again because they have not sought out the input that would lead them to ask some obvious questions);

    • Research on work-related violence has not yet made the connection in a widely accepted manner that moral or psychological and sexual harassment at work are forms of *psychological violence*. Simply put, we can say that these constitute *violence*;

    • Research on work-related violence has focused primarily on physical violence and its effects and costs without recognizing that psychological violence is equally damaging and costly to workers and employers;

    • This is a sensitive area because discussing and publishing about it widely means categorically that some managers should be removed from their positions;

    • Not enough trade unions, human resource professionals and top level managers are aware of the huge scope of the problem and the growing numbers of destructive managers found in the workplace;

    • There are legal issues involved, such as in harassment cases, where abusive managers will do anything to defend their actions and often are backed up by their own hierarchy.

23. "Nip it in the bud" means to catch something [negative] before it develops; to stop a negative development at its earliest stage, catching a potential problem in its bud phase before it becomes a fully "bloomed" destructive force.

24. "Setting the bar" refers to establishing the level of conditions and treatment in the workplace that you are willing to accept and tolerate. In the workplace, it is appropriate to set one's bar high, to set one's standards at a high level for the respect and treatment one expects and demands.

25. Top-down practices are those from government level, or top-management level in workplaces, while bottom-up practices are from the grassroots, generated by workers themselves in the workplace, or by community or other local-level groups.

Essentially, and eventually, top-down and bottom-up meet somewhere along the continuum. The meeting point often can become a jumping off place for discussion and discovery of the different points of view and needs of those at the proverbial top and those at the proverbial bottom.

26. M.-F. Hirigoyen, *Le Harcèlement Moral Dans la Vie Professionnelle: Démêler le Vrai du Faux*, La Découverte et Syros, Paris, 2001.

27. R. Semler, *Maverick: The Success Story Behind the World's Most Unusual Workplace*, Warner Books, New York, 1993.

28. J. Wislar, J. Richman, M. Fendrich, and J. Flaherty, Sexual Harassment, Generalized Workplace Abuse and Drinking Outcomes: The Role of Personality Vulnerability, *Journal of Drug Issues, 32*:4, pp. 1071-1088, 2002.

29. A number of seminal works on this issue include: O. Kernberg (ed.), *Narcissistic Personality Disorder*, W. B. Saunders, Philadelphia, 1989; V. Volkan, Review of Narcissistic Personality Disorder, *Journal of the American Psychoanalytic Association, 41*, 273-276, 1993; L. L. Altman, Case of Narcissistic Personality Disorder: Problem of Treatment, *International Journal of Psychoanalysis, 56*, pp. 187-196, 1975; J. Fernando, The Etiology of Narcissistic Personality Disorder, *The Psychoanalytic Study of the Child, 53*, pp. 141-158, 1998; A. Rothstein, Diagnostic Term: "Narcissistic Personality Disorder," *Journal of the American Psychoanalytic Association, 27*, pp. 893-912, 1979; M. Robbins, Narcissistic Personality: A Symbiotic Character Disorder, *International Journal of Psychoanalysis, 63*, pp. 457-474, 1982; O. Kernberg, Antisocial and Narcissistic Personality Disorders, in *Aggression in Personality Disorders and Perversion*, Yale University Press, New Haven, pp. 67-84, 1992; G. O. Gabbard, Personality Disorders: Narcissistic, in *Psychodynamic Psychiatry in Clinical Practice* (3rd Edition), American Psychiatric Press, Washington, D.C., pp. 462-490, 2000; S. Akhtar, Narcissistic Personality Disorder, in *Broken Structures: Severe Personality Disorders and Their Treatment*, Jason Aronson Inc., New Jersey, pp. 45-78, 1992.

30. *"Narcissistic pervert"* is the term accepted in the fields of psychiatry and psychoanalysis to describe this sort of destructive manager. The word "pervert" in the terminology appears to carry an emotional charge because of its sexual connotation. Since we are talking here about managers in the workplace, some people tend to immediately reject the concept, thinking that a term containing the word "pervert" could, therefore, not possibly apply to managers in the workplace. Rejecting the concept based on an emotional reaction to one word would defeat the purpose of this discussion. Therefore, throughout the rest of this chapter, *narcissistic pervert managers* will be referred to as *"dehumanizing managers"* to obviate use of the word "pervert." *Narcissistic pervert managers*, by definition, dehumanize their subordinates, are *dehumanizers*, and extremely destructive individuals. Such people are indeed perverse.

31. I. Nazare-Aga, *Les Manipulateurs Sont Parmi Nous: Qui Sont-ils ? Comment S'en Protéger?* Editions de l'Homme, Montréal, 1997.

32. M.-F. Hirigoyen, *Le Harcèlement Moral: La Violence Perverse au Quotidien*, La Découverte et Syros, Paris, 2000.

33. A. Soares, Comme 2+2=5: Le Harcèlement Psychologique Chez les Ingénieurs d'Hydro-Québec, La Prevention, *Performance, 19*, pp. 30-33, 2004.

34. In workplaces where people are not treated with respect, this scenario can unravel as if it were enacted straight from the psychology literature.

35. One should never underestimate the destructive nature of and capacity for perversion in a *dehumanizing manager*, even in an average, little *dehumanizer*. A *dehumanizing worker*, even one relatively low in rank and in the hierarchy, can bring an entire work unit to its knees. The destruction is calculated and takes place slowly over time.

36. P. Freire, *Pedagogy of the Oppressed*, Continuum, New York, 1970.

37. R. M. Sapolsky, *Why Zebras Don't Get Ulcers: An Updated Guide to Stress, Stress-Related Diseases, and Coping*, W. H. Freeman and Company, New York, p. 227, 1994 and 1998.

38. M. Kivimaki, M. Elovainio, and J. Vahtera, Workplace Bullying and Sickness Absence in Hospital Staff, *Occupational and Environmental Medicine, 57*, pp. 656-660, 2000.

39. It is noteworthy that there appears to be a growing crisis in the private sector in the United States, where CEOs remain on the job an average of some 18 months only. How can this be explained? *Dehumanizers* "sell" their images extremely effectively through seduction and manipulation so that they fill the ranks of these top and very important positions requiring true leadership capacity. They seduce with promises of righting the wrongs of their predecessor(s), of bringing fresh approaches to carry out the mandate of the company. Yet, one after the other, they get "found out" as being show, talk, and unable to fulfill the requirements of the job, unable to bring to realization the company's objectives. What is important for our purposes is that this repeat scenario points to the lack of training and awareness among those who recruit manipulators for such important jobs. Until one becomes aware of the characteristics to look for, one tends to remain in the dark, subject to the traps of *dehumanizers'* manipulations. Happily, however, once you know what to look for, awakening is rapid, and one's bar gets quickly raised. In the meantime, the costs are immeasurable—consider not only repeated headhunter costs, the time and investment of learning curves, and "wait periods" for a new CEO or senior manager to be able to carry out a company's mandate, but the costs to the people working in such companies cannot even be estimated. Change in "leadership" every 18 months brings distress and demoralization to many working in the affected company. Little momentum can be expected under such circumstances. In the affected companies, this scenario is surely a key reason for falling profits, reduced productivity, increased worker turnover, high absenteeism, high rates of depression and other mental health disorders, cardiovascular disease, and MSDs. To pick up the slack, management typically will look for ways to pressure the work force to work harder and faster.

40. N. Sokoloff, *Between Money and Love: The Dialectics of Women's Home and Market Work*, Praeger, New York, 1980.

41. J. V. Johnson, Collective Control: Strategies for Survival in the Workplace, *International Journal of Health Services, 19*, pp. 469-480, 1989.

42. Many examples of positive change to improve work conditions through collective action are described in M. Keith, J. Brophy, P. Kirby, and E. Rosskam, *Barefoot Research: A Workers' Manual for Organising on Work Security*, International Labor Office, Geneva, 2002, and in a multitude of trade union publications.

43. Psychology should perhaps play a greater role in choosing individuals for managerial positions where they have authority over subordinates. However, it should be noted

that the use of psychologists by companies or organizations has its risks for abuse. Psychologists are sometimes "bought" by top management to make decisions that are those of top management.

44. BKYD Management Consultancy, adapted from an Assessment Centre Project used in major companies and corporations in multiple countries, crossing different cultures.

45. Nor does psychotherapy. *Dehumanizers* may be among the "unanalyzable" since they are unable to question themselves; an analysis explained to me by a seasoned psychologist during my research for this chapter. Her practice consists mainly of patients whose lives and health have been destroyed by *dehumanizers*.

46. D. J-F. de Quervain, U. Fischbacher, V. Treyer, M. Schellhammer, U. Schnyder, A. Buck, and E. Fehr, The Neural Basis of Altruistic Punishment, *Science, 305,* 2004.

CHAPTER 6

Not So Radical Concepts to
Improve Workers' Health

MODERN DAY TAYLORISM

A turtle may live for hundreds of years because it is well protected by its shell, but it only moves forward when it sticks out its head [1].

An intricate relation exists between management practices, psycho-social factors, and work conditions as causes of MSDs and psychological distress in airport check-in workers. It is equally true for workers in other occupations. In seeking pro-active and preventive occupational health and management measures to improve motivation, worker participation is key. Recommendations and solutions to improve work conditions for both workers and managers are presented in this chapter.

Researchers only recently have undertaken studies to look at how various management methods and techniques, such as total quality management, team work, job enrichment, lean production, process re-engineering, just-in-time policies, and benchmarking or continuous improvement, may contribute as risk factors for negative health outcomes, particularly MSDs, among workers [2-4]. Investigations in this area are important to understand the perspectives of both employer and employee. For example, benchmarking and continuous improvement have been associated with impacts on workers' health in automobile assembly plants, where low control, low autonomy, plus high work demand, and poor work methods have been shown to cause ill health, fatigue, stress, MSDs, and dissatisfaction among workers [5].

Frederick Taylor was the person essentially responsible for the modern factory, where thousands of faceless workers perform relentless and repetitious tasks under tight supervision and control. Taylor divided complex manufacturing processes into small, individual tasks, convinced that this was the means of achieving maximum productivity. The beliefs and practices inherent in Taylorism are found in all kinds of work, workplaces, factories, companies and organizations today, an accusation most modern day managers would be loath to accept.

Today's organizational management practices are defended by industry as a departure from Taylorism and its inherent total control over workers, which has been seen as an absolute necessity for adequate management. For industry, work intensification policies, such as lean production and just-in-time policies, with quick turnaround policies dominating the air transport industry, are means of increasing productivity and generating productive gains. These management practices, it is claimed, are the answer to industry's difficulties in responding to changing customer demands and preferences, while enabling management to rely on the problem-solving ability of workers [6]. However, in reality, these practices have not demonstrated real success in promoting well-being among workers or diminishing work-related ill health [7].

It is erroneous to think that Taylorism is a relic of the past. On the contrary, the principles of Taylorism are completely engrained in modern day management practices. The roots of the modern day job description can be traced back to Mr. Taylor himself, limiting workers' potential and the possibility of job enrichment, with a direct negative effect on motivation. When is the last time you saw a job description that included an employee's aspirations, embodying what she wants to do, rather than something like a fixed list of limited tasks dictated to be performed? Human beings have a need to feel a sense of accomplishment in the work they perform, and most people want to be sure that what they make or do is necessary. They want to like making what they make or doing what they do. Work that is inherently Tayloristic in nature or dictated and performed under strict control is anachronistic, counterproductive, and contrary to creating a sense of accomplishment in one's work. Visiting most any "modern" day factory is a stark reminder of how little the world of work has advanced since Taylor. It is high time the world of work took a giant step forward.

The assembly-line way of working—individual tasks performed in a controlled amount of time in a repetitive manner and controlled by some level of supervision or mid-level management—has been transposed to a wide variety of service-sector jobs and white-collar office work. We need think only as far as supermarket check-out work, computer clerical work, hotel housekeeping work, or airport check-in work. Due to time elapsed, the introduction of new technology and the reactions against Taylorism that have taken place over the last several decades, one prefers to be seen as departing from Taylorism rather than applying it. But this idea is fallacious given what is practiced, and constitutes modern-day management mythology. The reality should be exposed better and discussed openly, since it is only through consciousness that agency is developed. Thus, writing broadly and discussing globally the impacts of modern management practices on workers' health is pivotal in catalyzing a change process.

Some little-known personality characteristics of Frederick Taylor himself are useful to understand the nature of work in modern times [9]. The man on whose guidance so many past and modern management practices have been built was an extreme obsessive-compulsive personality type, who even as a child counted his steps, measured the time for his various activities, and analyzed his motions in a search for "efficiency." Today such a child could be helped by medication or other therapy to improve his quality of life. Although called "Scientific Management," there was and is nothing scientific about Taylorism. Braverman elucidated the "success" of Taylorism—Taylor was simply lucky in his timing, for the worldwide growth of capitalist enterprises found rationality in an obsessive-compulsive-driven time/task/control sequenced way of working [10]. It is sobering to consider that the whole world adopted—on an unimaginable scale—the recommendations of an obsessive-compulsive neurotic crank. Surely, we could have done better.

Work intensification is often achieved by applying moral pressure on workers, a management technique used to increase the rhythm of work. Management practices designed for work intensification have been associated with intense psycho-social distress and strain among workers, who may fear to speak out critically about the system of work and the methods of work organization. One of the demonstrated results of such practices is a gap between how work tasks are prescribed and the reality of how they are performed. A variety of both psychological and physical adverse health effects result from such managerial practices [11]. In fact, the increase in work intensification can explain the fact that 50% of declared work-related diseases in France today are MSDs [12].

An increasing amount of research examines the impacts on workers, the environment, and society in general from the economic acceleration that is taking place in our times. In reality, the increase in work intensification together with new production methods and management practices leads to a deterioration of work conditions and a loss of autonomy for workers, rather than increased individual autonomy. The arguments used by industry in favor of the benefits to workers are not grounded in analyses of work conditions or on worker health impacts [3].

The evidence is stacking up showing that so-called "modern" management practices cause a worsening of work conditions, not an improvement, and that they directly lead to negative health outcomes among workers. To ensure public safety, study after study is required before consumable goods or pharmaceuticals are allowed on the market for mass distribution. Yet chemicals, tools, machines, work processes, and management practices are allowed into workplaces with absolutely no prior scrutiny. Observation and "study," if and when they take place at all, are post-facto when workplace practices and substances are concerned. And even when negative outcomes are demonstrated, it is usually very difficult to get changes introduced or bans mandated.

A more reasonable approach would be to apply a precautionary principle, requiring study and scrutiny of "new" management practices, chemicals, work processes, and so on, before they are introduced into workplaces, and then decide whether the evidence justifies widespread introduction. Such an approach likely would have a significant effect on the pharmaceutical industry. After all, working people constitute the largest single consumer group in the world, and poor work conditions are directly linked to consumer behaviors, particularly those related to pharmaceuticals and health services [13].

Particular to the group of management practices known to cause negative health effects are Taylorism, neo-Taylorism and Fordism, techno-bureaucratic management style, and management based on competitiveness, more or less the leading management practices used in today's world of work. These modes of production reduce opportunities for workers even to take breaks, while work intensification increases. Taylorism, neo-Taylorism, and Fordism are recognized today to cause MSDs, among a host of other adverse health outcomes, resulting from repetitive and monotonous task work, intense time pressure, a high level of work demand with low or no control by workers, and the absence of autonomy [12, 14].

Some research has used an historical analysis to demonstrate that while the practices employed in today's socio-economic system are more productive than any previous time in history, they are also more destructive than ever before [15]. Such findings are strong statements about the times we live in and what working people will put up with. Perhaps we should ask ourselves why we accept the unacceptable. Research associates present systems of production and organizational practices dominating today's socio-economic model with negative impacts on workers' health, with the associated costs routinely being externalized by companies. These findings should concern managers since necessarily production and performance cannot be at optimum levels when workers suffer, either in silence or with the knowledge of management, and when workers' energy, knowledge, and skills are not well used or not well focused [16]. If companies had to absorb all direct and indirect costs associated with work-related ill health, the very nature of production techniques might be radically different, radically improved.

Business examples demonstrate the many, many good things that can happen when the thin veil of Taylorism in modern day management practices is confronted and eliminated. Higher productivity and increased profits are just the beginning. For example, Semco, S.A., a Brazilian manufacturer of industrial equipment, achieved a growth rate of 600% and moved from 56th to 4th in its industry, in spite of periods of a faltering Brazilian economy. The example speaks to the success achieved by abolishing all practices based on Taylorism, doing away with the traditional corporate pyramid as an organizational structure, involving workers in all aspects of organizational decision-making based on a corporate culture of true democracy and complete transparency. A truly democratic workplace includes strictly defining the difference acceptable between top level salaries (including CEOs') and those of entry level workers. In the United States, a CEO's salary can be 400 times greater than that of a worker in the same company. Such gaping differentials áre in direct contradiction with the principles of democracy.

A worker at Semco, and member of the factory committee at one of the company's manufacturing plants, described in the following way the company's transformation to a transparent, participatory, democratic environment, "It was hard to get used to. You know, like when people are used to living in an authoritarian regime. When they finally let you out of jail, you can't believe it's true. The workers are motivated to work" [1, pp. 205-206]. Another employee reported, "The company became a paradise to work in. Nobody wants to leave" [1, pp. 206].

What of the jobs that managers perform themselves? Besides carrying out the ordained practices, the majority of jobs that "support" the labor process (such as personnel departments, administration, and management) merely make adjustments to existing systems without questioning or re-thinking the basic concepts of work processes or the organization and design of work. Management today has become largely administration, a process of labor conducted for the purpose of control within an enterprise or organization. Management itself is similar to the process of production, except that it produces no product other than the operation and coordination of the enterprise [10].

> Work itself is organized according to Taylorian principles, while personnel departments and academics have busied themselves with the selection, training, manipulation, pacification, and adjustment of "manpower" to suit the work processes so organized [10, p. 87].

Work can be a source of psychological and physical stability for a worker. Stimulating work can generate motivation and the work environment can have positive effects from a social perspective. In direct contrast to these positive effects, repetitive, high speed, low-skill work, de-humanized work tasks, using only a fraction of the worker's capacity, lacking in recognition for the contribution made by the work performed, can have traumatic effects for the worker performing such tasks. The effects of such work have been shown to result in a range of destructive psychological consequences, including feelings of humiliation, loss of self-esteem, and loss of motivation, with consequences for work performance and pernicious effects on life outside of work. A wide range of somatic effects has been documented as well, directly associated with the psycho-social factors in the workplace [17].

> Human nature demands recognition. Without it, people lose their sense of purpose and become dissatisfied, restless, and unproductive. Stalin understood this. Prisoners in his gulags were obliged to dig enormous holes in the snow, then fill them in. It broke their spirits [1, p. 109].

What we know about the relation between work and negative health outcomes points to management practices as both the direct and indirect causes. We cannot disassociate from working life the spillover effect into workers' personal lives caused by disempowering management practices [18]. To illustrate the association between direct and indirect causes of work-related ill health and their impacts on private life, important numbers of airport check-in workers reported that pain from MSDs interfered with their sleep and activities outside of work. As noted before, the trend in research on MSDs has been to focus on purely physical factors, while there is a need to focus on the multi-factorial etiology of MSDs and to pay special attention to the contribution of management practices in their development and severity.

Both individuals and organizations suffer when management practices are not conducive to enhancing the well-being of both. If organizations are sick due to management, then the people working there can be sick from certain styles of management [19].

Similarly, it is increasingly recognized that responsibility can be attributed to "careless management" for conditions under its control that contribute to work-related illnesses and accidents [20]. This relatively recent recognition contrasts with the age-old belief that worker carelessness is the cause of work-related accidents and injuries. Similar views have emphasized the fundamental need for organizations to invest in their human resources [15]. Investment in the human element of work necessarily yields positive outcomes for worker health and well-being.

So how may one envision making work more conducive to well-being? It is the participation of workers in organizational decision-making and enterprise activities that allows workers to go beyond the "minimalism" of work [15], creating the conditions where workers can be free to take initiatives and help organizations take risks, while having a buffer against the potential negative effects of the risk-taking. Semco has practiced true participatory management for decades. Equality in the workplace figures predominantly in Semco's company philosophy, and the results show very low worker turnover, decade after decade of profit, and continued growth even during severe economic downturns. But elsewhere, the non-involvement of workers in the conditions driving production diminishes the development and expression of talent that is needed in any organization.

> . . . [T]he era of using people as production tools is coming to an end. Participation is infinitely more complex to practice than conventional corporate unilateralism, just as democracy is much more cumbersome than dictatorship. But there will be few companies that can afford to ignore either of them [1, p. 107].

Promoting workers' well-being is also intricately linked to reducing workers' feelings of powerlessness, and indeed herein one cannot escape a political context. The feeling of powerlessness cannot be reduced without releasing some of the power held by those in top positions. Power is not an absolute; it is a perception, and it fluctuates in intensity, degree, and time. It can shift from one person or group to another, as do alliances. As one person relinquishes some of his power, another can pick up the slack. Reducing feelings of powerlessness cannot be accomplished by making changes in name only. Real changes require shifting the balance of power, genuinely allowing workers more decision-making latitude, autonomy, and control over the way they work. Here lies the crux of motivation.

Some researchers and business people [1, 12] promote participatory forms of management as perhaps the most health-promoting for workers at all levels, while not ignoring the problems and risks accompanying it as another form of managing people. Participatory management is contrasted with past and most present forms of management. If worker participation is to be a reality and to succeed in organizational decision-making and change, then worker autonomy is fundamental in order for individual creativity to express itself. The absence of autonomy among workers leaves no room for creativity, since management control and labor creativity are contradictory [3].

Consultative democracy, which, in most organizations means that top managers can speak their minds before the CEO makes his decision, is hardly democracy. Democracy in an organization means that all employees have a voice in corporate/organizational decisions—not just managers. So why is it the usual practice that one individual has the final say in most organizations and companies? Centralized power, as seductive and addictive as it may feel to that all-powerful individual, is highly risky. Having one person at the top doing as he wishes is seldom good for any company or organization. People at the top often lose their powers of perception and often make poor—sometimes disastrous—decisions. The inner drive to control more and more, and to be bigger and bigger is an immature urge and needs to be consciously controlled. Without any question, democracy is hard and constant work.

There is another important negative consequence of top-down-only approaches at the workplace. When management policies are determined by top management but applied only further down the hierarchy, worker health also tends to move down the hierarchy. Where designated responsibility does exist for worker health, it is often a function assigned to a technical unit or absorbed by a human resource function. Yet without proper information and training on the broad range of physiological, sociological, psychological, and management issues that can impact workers' health, how can the individual(s) given this responsibility realistically implement a pro-active approach to health in the workplace?

The protection and promotion of worker health must be an integral part of top management policies, with an established system incorporated and communicated at all levels of an organization, where the participation of workers is implicit. Organizations that give priority to health incorporate this function into their organizational strategy. Pro-active enterprises place monitoring functions at all levels of the organization. Worker health and safety, in this way, becomes everyone's business, and workers are enabled, encouraged, and empowered to participate in the implementation of related practices and policies.

The significant discrepancies found between employer injury reports compared with check-in worker self-reports, particularly on MSDs, demonstrate the lack of priority given to worker health by management of check-in workers. At three study sites, no organized method was found for addressing health-related issues. Worker health and work conditions would be best if fully integrated into management policies applied to check-in workers, established in consultation with workers.

The link between work organization, the management environment, women's work, and health outcomes has essential cost implications for management and workers due to lost work time and reduced performance on the job. Applying an integrated perspective is very much needed if we are to understand the different forces working against good health and well-being and to make changes for the better.

STRUCTURAL, ORGANIZATIONAL, AND EDUCATIONAL INTERVENTIONS TO REDUCE MSDs

Protective Strategies Can Reduce Risks

It is impossible to design jobs to fit 100% of the working population because workers vary in so many different ways. Gender-based factors as well as variations in size, strength, and age are some of the variables that make it difficult to accommodate all workers at every job. Adjustability and flexibility are, therefore, important criteria in job design. Research in ergonomics recommends that jobs be designed to fit 90% of workers with selection techniques used to screen out the remaining 10% [21]. Such practices can be questioned as discriminatory, in addition to not always being feasible. Designing jobs to fit 75% of the workers often is considered acceptable, although this level assumes that a percentage of workers may experience negative outcomes from a poor fit. In reality, many jobs are designed for the "average" worker, which should be considered unacceptable to managers, workers, and industrial designers, given that 50% of the population falls outside the mean. This means that all workers who are not "average" in size or shape have to adapt themselves to an ill-designed job, work process, or work station, increasing the likelihood of adverse outcomes over time for the workers. Better nutrition, increased access to clean water, improved maternal health, and the growing problem of obesity in many countries have meant that global trends are for people to be taller and bigger than ever before in many parts of the world. Why then are adjustability and flexibility not yet the standard for work station, work process, and tool design?

The magnitude of the problems, severity, and sheer numbers of workers affected by the frequency, duration, and intensity of exposure to poor work station and poor work process design led to significant research undertaken by the U.S. National Academy of Sciences. The research showed a clear relationship between back disorders and physical load, including manual material handling, load on the body, frequent bending and twisting, and heavy physical work. Covering all industrial sectors, work-related psycho-social factors associated with low back disorders were identified as rapid work pace, monotonous work, low job satisfaction, low decision latitude, and job stress. High job demands and high job stress were identified as work-related factors associated with upper extremity disorders. Results of the Academy's research justified the need for a national ergonomics standard in the United States, which was subsequently put in place under the Clinton Administration but immediately revoked during the Bush Administration's first term. Today, no national ergonomics standard exists in the United States, leaving workers no legal framework to fall back on other than OSHA's General Duty Clause, use of which has been associated with serious limitations [22]. This example emphasizes the importance of collective voice, and of the limitations of reliance solely on top-down systems to protect workers. Even when the evidence is stacked up, politics may play out a hand that does not favor working people.

The scientific evidence and industry data strongly indicate that properly implemented strategies to reduce the incidence, severity, and consequences of work-related MSDs can be effective. Successful programs can be found in a variety of job settings, taking into account work procedures, equipment design, and characteristics specific to the organization, usually involving a high level of commitment from employers and workers [22, p. 365]. Job re-design, in particular, is found to be an effective means of preventing, reducing, or eliminating MSDs in industry [23]. However, as job re-design improves work stations and work processes, work organization and psycho-social risks increase in importance.

Recognizing similar necessary factors for successful workplace improvement, the tripartite European Union Advisory Committee on Safety, Hygiene and Health at Work (also known as "the Luxemburg Committee") adopted in May 2001 a final draft calling for consideration of a new regulatory initiative (such as an EU Directive) focused on repetitive and monotonous work. The requested initiative recognizes that work-related MSDs can be prevented when the issue is approached in a multidisciplinary and participatory manner. It calls for a number of necessary elements to ensure the success of workplace interventions, including:

- Commitment on the part of management;
- Involvement of workers;
- Assessment of workplace risks;

- Identification and implementation of corrective and preventive measures;
- Training and information—for both employers and workers;
- Appropriate medical management [24].

A convincing body of evidence shows that work-related MSDs can be minimized through structural, organizational and educational interventions [25]. Structural measures optimize work station design and equipment. Organizational measures focus on job design, including the distribution of tasks, speed of work, and rest breaks. Educational and training programs should target both workers and managers for disseminating information, generating ideas, stimulating innovation, sharing solutions, preventing problems, raising awareness, and validating individual knowledge and experience. This three-pronged approach is an effective multidisciplinary, holistic strategy to use in addressing airport check-in work so as to prevent the onset of work-related MSDs. Managers often are reluctant to re-design jobs because of costs. But committing capital to the re-design of jobs where needed often can be cost-effective, shown by decreases over time in medical and compensation costs, and increases in worker motivation and therefore performance [21, 22, 26, 27].

Similar multi-level prevention strategies are demonstrated as effective in regulating and reducing work environments associated with psychological distress and job strain. Tackling MSDs requires addressing simultaneously those factors known to induce negative stress, while tackling work-related stress requires dealing with the factors causing MSDs as well. Historically, the most successful prevention strategies have used a multi-level approach of participatory interventions, local organizational change involving grassroot organizations and trade unions, and national legislation providing a legal framework for worker empowerment. Globalization and increasing international competition have made the climate for stress prevention less favorable today, suggesting a need for a renewed multi-level prevention strategy combining scientific investigation, local-level democratization, and political mobilization at societal and global levels [28], including legislation to address (from a top-down perspective) occupational stress [29].

Voluntary agreements are an alternative to national legislation as part of social policy; one example is the "Stress at Work" Code, signed by employers and unions at the European regional level in 2004 as a pledge to combat stress at work. The code is designed to increase awareness of the problem and encourage companies to reduce and eliminate the stress or prevent its root causes. Responsibility rests with the employer, but measures should be carried out with the participation of workers. The code notes that stress can be triggered by any number of factors, from abusive behavior at work, noise and exposure to dangerous substances, social pressure, a perceived lack of support, or just a feeling of inability to cope. While voluntary agreements may show good will, how they are implemented and followed-up remains debatable.

It has been suggested that organizational health is made up of both organizational performance and worker health and satisfaction outcomes, and that a balanced approach to maintaining worker well-being gives comparable weight to both. There is a clear link between a firm or organization's performance and progressive approaches to human resources management [30]. The World Health Organization, under its rubric of Healthy Work Organization, has promulgated a similar concept aimed at maximizing the integration of goals for worker well-being and company objectives for profitability and productivity, a means of creating a balanced scorecard for enterprises. The integration of these goals and objectives is important in minimizing negative health effects of occupational stress. Indeed, not involving workers in the planning of work stations, work organization, work processes, and workplaces overall is "an outdated management philosophy that aims to control nearly everything from the top down and ignores the know-how and needs of people who create goods and services" [31].

Using "Voice" to Improve Performance

A company's performance, productivity, and effectiveness are influenced by a variety of organizational practices, including worker empowerment, worker voice mechanisms, information-sharing with workers, problem-solving teams involving workers, and job re-design. A productive enterprise or organization affords workers growth, creativity, and change in the context of change in work organization and technology. Organizational-level outcomes, however, often overshadow worker needs and outcomes.

> Today's workers know more and have more than ever before . . . workers want to have some say in what they do at work, how, and when; they want to get meaningful feedback on their efforts; they want to be coached, rather than bossed; they want to be informed about developments affecting their jobs and future—in short, they want to enjoy informal participation at every organizational level [32].

The association between empowerment, collective voice, and health has been described eloquently by Johnson (1997), including an analysis from the perspectives of social justice and political and health movements' focus on the effects of degrading and alienating work processes [33]. Sargent and Terry also have demonstrated the beneficial effects of worker empowerment on health, job satisfaction, and productivity [34]. An empowerment perspective potentially changes hierarchical structures by stipulating that the "grassroots" of an organization should exercise decision-making, authority, and occupational self-determination. Low levels of empowerment in day-to-day work life have been associated with ill health, drug and alcohol abuse, mental distress, and risk of early death.

Research on social class differentials and health punctuates the inequalities and power differentials between those in the upper class and workers in lower socio-economic groups. The body of evidence on the social gradient of mortality and morbidity strongly indicates better health outcomes among those in higher socio-economic groups, and those in power-holding groups; workers in lower socio-economic groups are more exposed to work-induced stress likely to produce adverse health outcomes [35]. The evidence has mounted showing that workers have greater difficulty protecting themselves in countries characterized by wide income disparity, or where there are high levels of labor market, employ- ment and income insecurity. Countries typified by great income disparities and severe social and economic insecurity also tend to have weaker collective voice representation security [36], which is part and parcel of the difficulties that workers face in securing their own protections. Of no less significance and adding to what amounts to a double-edged sword, the magnitude of the effects on health produced by adverse work conditions may be higher in lower status groups due to their increased vulnerability [37, 38]. Airport check-in workers fall into this category of workers.

To sum up this "double-edged sword": workers in countries with wide income differentials, or in countries with high economic insecurity (i.e., low labor market, employment, income, or voice representation security) are more likely to be without work-related health protection than workers in countries having greater economic security or less income disparity. Compared with workers in countries enjoying greater economic security, those same workers are also more likely to have worse overall health measures due to poverty- induced poor baseline health compounded by occupational exposures. One could perhaps encapsulate this as the "occupational health factor of the culture of poverty" [39].

It is useful to consider the underlying philosophy of empowerment and work, which is, in fact, human liberation. Empowerment as a workplace concept has been introduced to refer to a change process that nominally increases the power of a group to influence an aspect of their existence. In direct opposition are so-called "modern" management practices, such as lean production, which often is accom- panied by downsizing and work intensification while claiming to increase rank-and-file worker participation in task-level decision-making, supervision, and job changes. Such practices equally invade white-collar workplaces under the guise of "employee engagement," and other euphemisms. These modern-day versions of Taylorism operate under a policy of "management by stress." In some workplaces under these management practices, individual workers may have increased horizontal control over tasks to some degree, however, this "gain" often is at the expense of little or even decreasing vertical control in decision-making at the organization [33].

In contrast is a democratization perspective, developed in Scandinavia, which emphasizes a multi-level strategy integrating workplace and national efforts to change workplace, legal, and administrative structures. This perspective has been attacked as "unrealistic" and not "productive" enough.

Management at airports around the world would do well to remain open to experimentation with various models or approaches to arrive at a system of worker participation that functions in each particular economic, political, social, and cultural context. Just as research results support the need for flexibility in work design and work organization to reduce MSDs, so too flexibility of approach is needed in managerial attitudes if full participation of the work force in organizational decision-making is considered a priority.

INCREASED WORKER AUTONOMY

Almost all businessmen think their employees are involved in the firm and are its greatest asset.

Almost all employees think they are given too little attention and respect, and cannot say what they really think.

How is it possible to reconcile these two positions?

The sad truth is employees of modern corporations have little reason to feel satisfied, much less fulfilled. Companies do not have the time or the interest to listen to them, and lack the resources or the inclination to train them for advancement. These companies make a series of demands, for which they compensate employees with salaries that are often considered inadequate. Moreover, companies tend to be implacable in dismissing workers when they start to age or go through a temporary drop in performance, and send people into retirement earlier than they want, leaving them with the feeling they could have contributed much more had someone just asked [1, p. 107].

A consistent association has been established between lower injury rates and worker empowerment, among other factors such as active management in health and safety. Lower injury rates have been found where workers are expected by management to use their own initiative, rather than carry out tasks based on management instructions, where there is encouragement by management of a long-term commitment by workers, and where management-labor relations are good. In addition, companies which provide longer periods of safety training and continued safety training regularly and companies whose workers have longer seniority show lower injury rates and lower staff turnover [40].

These findings demonstrate the positive results for both workers and management when management exhibits a positive attitude towards its work force. Airport check-in workers are a group with a high level of job demand but with low or no control over the way work is carried out. Their participation in decision-making is not built into the job and encouragement to take initiative is not part of initial job training, as with most low-skilled occupations. Support from management is absolutely necessary for organizational change to succeed in improving work conditions. However, "simply imploring, rewarding or forcing people to participate is not sufficient" [41]. For positive attitudes and approaches to be adopted more widely, a new "mind set" may be required and allowed to develop over time. Gradual acceptance by the work force may follow, and changes in attitudes among managers and workers may come about very slowly, not created by decree [32; 40, p. 215].

In Scandinavian countries, worker involvement in workplace decision-making is long established. In their model, formalized management systems are developing, since they are needed to give the work environment a stronger status in organizational decision-making. Workplace assessments are useful in initiating innovations and problem solving, but need to be supported by education and training for both management and workers [42]. The kind of assessments that can stimulate change require:

- *Enhanced worker motivation and job satisfaction:* Worker motivation and job satisfaction are increased by opportunities to participate in decisions affecting one's own work methods, everyday job routines, and performance goals. Having control over one's own work enhances commitment and quality effort.
- *Problem-solving capabilities:* Making use of workers' personal experience can be beneficial in decision-making in the organization.
- *Greater acceptance of change:* Involving workers in decision-making about organizational change can reduce resistance to change.
- *Greater knowledge of work and the enterprise:* Involving workers in problem-solving and decision-making increases knowledge of their own job and how it relates to the entire enterprise or organization.

Because the lack of control by workers and the concurrent exercise of management control over workers has been associated with negative health outcomes [3, 43, 44], increasing real opportunities for workers to exercise control over their work environment appears to increase worker autonomy. Increasing control over the environment also diminishes the potential for negative health outcomes associated with low levels of control over one's work environment [43]. The main challenge for management therefore would be to design jobs to be more intrinsically satisfying and motivating for workers to perform.

The combination of little or no ability to participate in decision-making together with a high level of job demand is demonstrated to constitute the highest in psycho-social health risk [14]. This categorization is shared by machine-paced operators, low-status service-sector workers (including waiters and cooks), many clerical jobs, as well as airport check-in workers, and is much more common among women workers than men. Unlike many "male jobs," high-demand jobs performed by women usually appear in combination with low decision latitude. Indeed it is constraints on decision-making, rather than decision-making itself, which present the greatest problem for most jobs [45].

Complementing the demand-control framework is the effort-reward imbalance, which is rooted in distributive justice [37]. Assuming that effort at work is spent as part of a contract based on the norm of social reciprocity, and that rewards are provided in terms of wages, esteem, and career opportunities, then balance is potentially achievable between the two. However, the lack of worker control over the jobs they perform coupled with the lack of autonomy and decision-making latitude introduce imbalance into the equation, which the rewards do not necessarily compensate. Furthermore, effort in the form of motivation, productivity decrease, and illness through distress-induced depression, chronic frustration, and dissatisfaction are common outcomes.

Low control over job tasks has been linked with a high degree of muscle tension, leading to MSDs [14, 43]. The high-demand and low decision-making latitude nature of airport check-in work appears to favor MSDs. The mechanism by which high strain jobs favor MSDs is not clear. It may be as simple as the fact that workers having a high degree of control over their jobs can better control their working postures (e.g. sitting down, getting up, changing the height of working surfaces, etc.). However, it would appear equally logical that a high degree of strain on a continuous basis promotes tension in the musculoskeletal system, favoring the eventual development of a chronic condition.

The call to management is to try to discover ways of increasing worker autonomy and control over the work environment as a means of improving health and well-being. The greatest challenge in undertaking this process may be for managers to recognize and accept that empowering workers through increased job autonomy and less control over them entails a break with the prevailing management paradigm that is focused on control, order, and compliance [16]. As stated earlier, this means empowering workers while disempowering those exercising control. It would perhaps be useful for managers to put this into context with the reality of the daily balancing act practiced by most workers, whereby workers' normal state of being is managing those workplace factors which may cause ill health and managing their own psychological defenses to these risks. For many workers, carrying out the daily balancing act of trying to not fall victim to the effects of ill-adapted work conditions, ill-conceived management practices, and abusive managers is highly stressful, a response which can lead to adverse health outcomes over time [18, 46].

Worker control has been defined as the worker's ability to control his or her own activities and not as control over others [45]. While this definition is consistent with a health model that focuses on the level of the individual, it may be problematic for the many managers, trade unionists, organizational designers, and industrial sociologists who address policies for macro level control at either the level of the work group or the organization.

> . . . organization policies may determine the possibility of worker partici-
> pation in a broad range of company level decisions. This possibility of
> participation can contribute to productivity and worker well being (such as
> reduction in feelings of powerlessness). Union membership and member-
> ship in influential work groups must also affect the total picture of decision
> latitude available to the worker. These macro social factors may have direct
> effects on health and productivity as well as indirect effects through
> changes in the possibility of task control at the individual level. Factors
> similar to these have been found to contribute to psychological well being
> [45, p. 141].

Evidence suggests that positive management support would result in fewer MSDs, and even have a protective effect against MSDs by acting as a buffer against physical and psycho-social risk factors [47]. The issue of control may be the critical factor directly causing negative reactions in workers facing job strains [48-50].

Understanding the importance of worker control in relation to job demands and decision latitude, it stands to reason that worker satisfaction and well-being would be improved by ensuring worker input in the planning process when new technology is introduced in the workplace. Computerizing or adding new technology to a job is an example of a practice typically carried out without the participation of workers in decision-making and which usually contributes to minimalizing a job. Computers and other forms of new technology are usually introduced as a means of "better carrying out operational policies." The intro-duction of new technology, however, does not necessarily reduce work pressure that existed prior to computerization or technology transfer due to repetitive and monotonous work. Integrating computer work into a job that was already Tayloristic in nature often makes the job even more painful and alienated from the worker's control, rather than the reverse [12]. For airport check-in workers and supermarket check-out workers, the integration of computerization into the job has added new unforeseen physical and mental health risks for the workers performing those jobs [51-55].

Statham and Bravo have provided supporting evidence of the association between worker control as a determinant of stress where job satisfaction is affected by worker input into decision-making related to the introduction of new technology [50]. Examining multiple organizations with traditional top-down control structures, where employee innovation was not encouraged, workers had no participation or involvement in planning processes and reported that management ignored their input. The introduction of new technology in the absence of worker participation was associated with adverse health effects, including feelings of anger by workers. Korunka et al. provide similar evidence; they found that introducing new technology to workers performing monotonous jobs (which are by nature low in control), such as computer clerical work, significantly reduces job satisfaction and increases MSDs and psychosomatic complaints [56]. Yet higher levels of worker participation result in lower levels of psychosomatic complaints and in increased job satisfaction, and working with new technologies was found to even enhance these positive effects, particularly due to the participation aspects. Lowe et al. show that technological change coupled with work intensification and re-organization contribute to an increased incidence of MSDs [57].

Work conditions for those in monotonous manual employment (predominantly women) can be improved by simultaneously advancing ergonomics, management styles, and specific tasks. Negative outcomes could be prevented easily with careful planning and feedback mechanisms when changes are introduced, not limited to the introduction of new technology. And of no less importance is the need for effective communication channels—in all directions of the organizational structure.

The creation of over-inflated expectations among customers, which workers may not be able to satisfy at the check-in, appears to contribute to psycho-social stress among check-in workers. Passenger expectations are created by management practices. For example, advertising and marketing strategies are used to promise that passengers always obtain what they want, when in reality often such promises cannot be fulfilled by the airline. Check-in workers are the first in line to meet with customers' reactions resulting from unmet expectations. "The check-in worker kept us there waiting for half an hour. She was questioning my wife's visa with our daughter's name on it. I got so angry I was ready to hit that woman." (Passenger in Amsterdam International Airport) Workers are confronted with passengers' complaints, dissatisfaction, disappointment, anger, and even violent reactions. Management practices known to lead to difficult situations with customers should, at the very least, be accompanied by mechanisms to help employees. Creating customer expectations without worker training to deal with the unexpected is disempowering for workers and contributes to high levels of stress on the job.

The abuse of positions of hierarchy is closely related to the lack of worker control in the workplace. Sapolsky describes with brilliance and humor the relation between social rank, individual control, the stress response, and disease outcomes, in both primates and humans [58]. In providing the "punch line" regarding the primate-human difference, Sapolsky punctuates the fact that humans have invented ways of subjugating low-ranking members of society (indeed, workplaces count as mini-societies, rife with social ranking) like nothing ever before seen in the primate world [58, p. 308]. Clearly this is not a distinction humans should be proud to hold. The importance of social support cannot be overestimated, and apparently explains, in large part, why zebras do not get ulcers while humans do. Understandably, wearing stripes will not solve our problems in the workplace.

> If baboons are spending only four hours a day filling their stomachs, that leaves them with eight hours a day to be vile to one another. Social competition, coalitions forming to gang up on other animals, big males in bad moods beating up on someone smaller, snide gestures behind someone's back—just like us. . . . They [baboons] have the same luxury [as humans] to make each other sick with social and psychological stressors. Of course, like ours, it is a world filled with affiliation, friendships, relatives who support one another; but it is a viciously competitive society as well [58, pp. 265-266].

Abuse of workplace hierarchy is a practice used to subjugate some workers. Those in positions lacking control, autonomy, and decision latitude are the most vulnerable to such abuse. The lack of control and autonomy and the exposure to abuses of power are strong predictors of stress outcomes, and not surprisingly often result in physical and psychological ill health. Management practices that exclude workers from decision-making constitute an abuse of workplace hierarchy.

EXAMINING JOBS HOLISTICALLY

Given the multitude of work-related factors that can affect workers' health and well-being, it seems absolutely necessary to examine worker health from a broad perspective. Doing so can lead to a basis for reducing job stress and related adverse health outcomes by examining ways to "balance" various elements of the work system, while recognizing that work conditions impose both physiological and psychological loads on the individual. The physiological load produces stress if it exceeds the available physical resources. But work conditions also can impose psychological loads that can have emotional, behavioral, and physical consequences. With understanding of this fact, it is entirely inadequate to examine only one element constituting workload when considering work-related stress.

One approach to reducing job stress is to integrate both physiological and psychological factors of work and person, a basis for balancing various elements of the work system. A useful example of this approach is found in the Smith and Sainfort "balance model" [59], which takes into account the individual, technologies used, tasks performed, the physical and social work environment, and the organizational structure. This approach can help to clarify the relationships between job demands, job design factors and loads, and the various elements interacting to determine the way in which work is performed, and to what degree the work is effective in achieving individual and organizational needs and goals.

Because psychological loads at work can have emotional, behavioral, and physical consequences, particularly in the case of stress-induced health effects, objective work conditions are not alone in contributing to adverse physical and mental health outcomes in workers. An individual's perception of the work conditions and the resources (psychological and behavioral) that the individual possesses to deal with those conditions are important determinants of the consequences. The perception of a job or task as monotonous, repetitive, or low in demand can create negative stress, resulting in adverse health effects. Research by Lowe et al. showed that communication at work, support, job demands, resources, and rewards were significantly associated with employees' perceptions or beliefs about whether their workplaces were healthful [57]. The evidence supports the idea that prominent concerns of employees are strongly related to workers' perceptions of healthful workplaces and has implications for worker commitment and morale.

Physiological and psychological loads interact and may even reinforce each other, essentially pointing to the fact that human health is the product of any number of influencing factors working in combination. Human physical, psychological, and behavioral resources change over time and are affected by both stress responses and work conditions, thus it is important to examine all of the influencing factors, not just one. In pursuing interventions to improve work conditions, it is necessary to examine both objective and subjective factors, including workers' perceptions of their work.

For airport check-in workers, these elements vary by the level of check-in system mechanization, but as demonstrated, a gamut of factors are linked to the outcome of MSDs. Physiological stresses—namely, job factors that involve lifting, carrying, awkward and static postures, aggravating ambient factors, frequency and intensity of exposure time—lead to MSDs. Similarly, psycho-social stresses lead to MSDs caused by stress-inducing factors that include passengers' emotional demands, psychological harassment by management, violence, aggressive passengers, passengers pushing workers to work faster, quick turnaround management practices, low-wage work with high demands, and lack of involvement of workers in the organization of their work, poor or no communication with management, and employment and income insecurity. Both management practices and the work environment thus induce or aggravate different forms of stress, leading to MSDs.

A growing number of researchers and scholars are demonstrating the critical role of management involvement in and commitment to occupational health and safety, as well as the impact of management practices on workers' physical and psycho-social health. An early researcher on job characteristics and worker well-being and a proponent of the view that the domain of work conditions should not rest in the hands of employers, Gardell emphasized the need to examine those job characteristics that provide worker satisfaction and well-being and to use those characteristics as foundation building blocks [60]. The key characteristics include control over work organization, work methods and speed, rhythm or intensity of work; understanding of the work process from a holistic perspective; development of and use of workers' technical and social skills; possibility for social interaction; and work scheduling.

A number of prominent researchers in the fields of occupational health, management, and ergonomics [61] describe the need to analyze work environments from a holistic or multi-factorial systems approach and to grapple with improving work conditions. In looking at the work environment from a holistic perspective, any one element will influence any other element. This fact makes it necessary to examine the sum of the parts in addition to the individual contributing factors. Systems-approach researchers bring together different schools of thought on the common grounds of worker autonomy, worker participation in organizational decision-making, and the need for job enrichment as means of increasing productivity, efficiency, and improving work conditions. These researchers, coming from different schools of thought, all recognize the pivotal role that managers play in an organization's approach to worker health.

Managers should be interested in and concerned with stress-reduction objectives since organizational goals of job enlargement, job enrichment, increased worker autonomy, learning, and capacity building are inhibited by high stress levels caused by any number of factors. A fairly substantial body of literature looks particularly at the impact of management methods on worker stress and associated adverse health outcomes. Scandinavian and North American schools of thought emphasize the structural or work organization sources of work-related stress [12, 43, 62].

Five core job characteristics have been linked to occupational stress, particularly related to job enrichment [63]. Skill variety, task identity, task significance, autonomy, and feedback are critical for job design, and various outcomes such as motivation and performance, while critical job elements that are coupled with individual characteristics determine the success of job enrichment. The ability of the worker to make a psychological connection with the job defines the influence of the job on the person, which has a direct relation to the experience of work-induced stress.

These consequences are determined primarily by the individual's perception of the work conditions and their available psychological and behavioral resources, but not by the actual physiological load. A poorly designed chair puts physiological load on the worker's musculoskeletal system and may produce an MSD. The chair also may be perceived as uncomfortable and therefore represent a psychological load, which may lead to stress responses. Physical, psychological, and behavioral resources change over time and are influenced by both stress responses and work conditions.

Individual technologies used, tasks performed, physical and social work environment, and the organizational structure all interact to determine the way in which work is performed, and the effectiveness of the work in achieving individual and organizational needs and goals. These elements work in concert to provide an individual with resources to achieve his and the organization's goals. Job stress is a potential negative attribute of each element, but positive aspects of each element also can counteract negative influences from any of the other elements. It may sound rather vague and academic, but the approach can be effectively applied in daily work without difficulty.

> . . . the negative influences of inadequate skill to use new technology can be offset by worker training. Or the adverse influences of low job content can be balanced by an organizational supervisory structure that promotes employee involvement and control over tasks. Proper job design can be achieved by providing those characteristics of each element that can meet recognized criteria for worker ego needs fulfillment and that set proper physiological and psychological loads. However, as such a perfect job cannot often be achieved, then using elements to compensate for poor aspects in other elements can balance the stress with moderating factors that reduce the load and potential health consequences [59, p. 76].

This approach could be used to reduce or eliminate MSDs as well as job stress. When balance cannot be reached by changing the negative elements of the job, then improvement may be achieved by enhancing the positive elements of the job. In this way, the positive aspects of work can be used in an attempt to counterbalance the negative aspects, with the goal of preventing MSDs. Similar findings have been shown by Sargent and Terry, with evidence that high levels of task control can act as successful buffers for the effects of job demands [34]. Of equal importance are the findings that task control exercises a significant effect on job satisfaction [34, p. 219].

We naturally counterbalance various aspects of work, using our defense mechanisms to ward off negative psychological impact from poorly adapted systems of work. This natural counterbalancing, or adapting oneself, may explain, in part, the low rate of absenteeism among airport check-in workers, despite severe MSD pain and aggravating work conditions.

This approach has been mentioned as an example of how the design of jobs can be looked at from a holistic perspective, emphasizing the job's positive elements that can be used to overcome, at least to some degree, negative aspects. And it can be done without focusing on any single factor, such as shift work, or demand and control. All aspects of the job, therefore, must be considered in developing a proper design, which cannot be accomplished without involving workers. This approach does not subscribe to one single approach for job design, such as content enrichment or participation. Rather, it is similar to an organizational development approach, which attempts to solve specific problems using one or more aspects from any number of perspectives. Stress, not productivity, is the outcome of interest, and therefore may direct different interventions than would an organizational development approach.

The approach could be expanded to help analyze and define causes (and cures) of occupational disease, beyond work-induced stress. The concept could be applied to a variety of jobs, in a variety of cultural contexts, including airport check-in work, supermarket check-out work, airport baggage-handling work, or to a variety of computer-based jobs. The concept of "load," for example, could be used to analyze both cumulative load on the body and the actual weight of a load. In the case of airport check-in workers, manual handling is influenced by how "load" is perceived, not simply by the actual weight of baggage.

An emphasis on ergonomic aspects of work and the participation of workers in work organization and work design can meet one another through efforts at balancing and compensating factors related to each of these two areas. For example, highlighting the positive factors to counterbalance the negative factors that cannot be changed is one means of making immediate improvements in work conditions. This is similar to applying simple ergonomic solutions to improve work conditions, which has been used successfully in developing countries [64].

The emotional strain associated with service-sector work performed by women [65] combined with a high level of job demand with low decision latitude fills out a picture of airport check-in work as a potentially highly stressful job. These objective psycho-social factors, associated with the very nature of the job, plus dealing with the emotional demands of passengers, and the applied management practices are supposedly significant causal factors in MSDs. Such factors should be considered regardless of whether a check-in system is semi- or fully mechanized.

Regular pauses at naturally occurring intervals (meaning pauses decided by workers themselves) have been shown to reduce the effects of static posture and prevent MSDs [66]. Similar benefits are achieved through job enlargement, decided with worker participation, where workers can perform a variety of tasks. Not surprisingly, preventing natural breaks has been shown to result in a long-term accumulation of strain symptoms and a long-term deterioration in work capabilities. To reduce the effects of static posture and help prevent MSDs, airport check-in workers require regular breaks, similar to office-based computer workers and supermarket check-out workers.

From a management point of view, a long-term cost-benefit vision that considers factors affecting workers' long-term commitment to productivity, quality of service, worker commitment to job and employer, and worker well-being is more beneficial to both workers and employers than a short-term vision. Without this vision, it would appear that some enterprises could more appropriately be conducting "savings analyses" related to worker health, rather than "cost analyses." And who is better placed to shed the light where it is needed than those performing the jobs themselves?

PROBLEMS AND SOLUTIONS: WORKERS KNOW BEST

The involvement of workers in identifying and solving their own problems in the workplace was explored in the educational pedagogy of Paulo Freire. Writing in the 1970s about the intimate knowledge that adults have of their own work and life situations, Freire was a strong proponent of the learning and problem-solving that can be accomplished by drawing out that intimate knowledge. Drawing out, validating, and making use of adults' real-life knowledge and experience can be transformative, empowering to a point where collective problem-solving and action can be initiated [67, 68].

The work of a variety of researchers, scholars, academics and practitioners is grounded in an understanding of the benefits of involving workers in research [69], but, unfortunately, such work is still today in the minority. Adults learn best through images and concepts that relate to their everyday reality; they know best what their problems are and collectively once encouraged with their own knowledge validated, usually generate solutions applicable to their life situations. In a workplace context, workers are the experts in the jobs they perform. They know intimately how the job affects them, but their knowledge and experience often needs to be actively brought out and validated. Workers are an important source of information and they are more likely to "buy into" solutions when they are involved in the process [21, 67, 68]. To facilitate workplace learning, versus teaching workers, and to create lasting change, workers need to be involved in the process of identifying and solving problems.

Worker participation, effective use of worker voice, participatory ergonomics, or establishing changes in work conditions through participatory approaches involving both workers and managers, are demonstrated as central to reducing MSDs and improving overall workplace health and safety [24, 64, 42, 70-74].

In its broadest sense, worker participation has been defined as " . . . any limitation set by the workers on managerial discretion to make and implement decisions. Thus, it covers all forms of worker participation in decision-making" [32, p. 385]. In organizational decision-making, worker participation can be formal, using institutionalized mechanisms, or it can be through informal approaches. Informal approaches include agreements between trade unions and employers and comprise elements such as job enrichment, management by objectives, democratic leadership, autonomous groups, and flexible work hours [32, pp. 386, 392].

At a practical level, worker participation in decision-making refers to the degree to which workers have opportunities to control or influence their job environment, specifically relating to work conditions [75]. Well-designed jobs stimulate and reduce fatigue by creating variety and challenge and eliminating monotony. Jobs with worker participation built in create more productivity and higher efficiency, providing benefits to employers. Increased efficiency through worker participation can be seen as a result of making use of the knowledge, initiative and creativity of all personnel, ". . . and not only of a managing elite" [32, p. 390].

There is an important gap between the prescribed organization of work and the way work is actually carried out. Workers demonstrate adaptive abilities every day, responding to changing demands in the workplace, finding solutions to problems as they arise. But managers are generally unaware of this disparity, largely due to a lack of organizational commitment to worker participation in decision-making or work process design. Management practices using only top-down decision-making that excludes workers has been called "fraudulent" decision-making [46].

Increasing participation in the workplace increases job satisfaction, improves decision-making, and generates important cost savings for employers. Worker participation programs in the United States, for example, were found to make jobs more satisfying and in the process saved organizations $2.2 billion in 1988 [76]. Work re-design implemented with the participation of workers results in better work performance and fewer work-related injuries and illnesses, including negative psycho-social outcomes [14, 76-80].

For organizations that involve workers in decision-making, long-term benefits are associated with continuous improvements in the work environment. Once workers are activated to participate, then initiative, which is in part an outgrowth of autonomy and reduced feelings of powerlessness, generally continues as long as there is management commitment and support. Involving workers in workplace design is an essential challenge for management in contemporary workplaces [31, 81]. The very process of involving workers should make people more aware of the settings around them, inspire them to ask themselves what they are trying to do there, stimulate them to assess how appropriate their settings are for what they want to accomplish, and lead to the changes needed to provide a better fit between workers and the setting [82]. Workers must, therefore, be personally involved in developing and implementing technology from the design stage. Where technology already is developed, participation should be in assessment, evaluation, and design of strategies for problem-solving and work optimization.

With all of these near-guarantees for success, how should we explain the limited success of worker participation? One had best turn to management scholars for the answer: attitudes, emotions, fear (nothing that cannot be dealt with, but indeed it requires work). Managers mostly have an inherent lack of trust in workers' qualifications and motivations. Managers typically fear losing control, and workers typically fear being co-opted by management [32]. Notwithstanding, many positive examples exist where worker participation has contributed to improving work conditions and increasing worker well-being. (Despite an encouraging body of literature, however, there are far more examples where managers' attitudes have not yet changed than there are of the contrary.)

No single principle will solve all problems in any one industry. In general though, using tools based on a multidisciplinary problem-solving approach is powerful and cost-effective. Organizations do well to remember that using simple tools (such as examining a worker's body, studying its capabilities and limitations, and then applying this knowledge to the design of work) is not about sick workers, rather it is about sick jobs, and how to improve ill-designed jobs and reform destructive management practices as a means of eliminating or reducing accidents, MSDs, fatigue, discomfort, boredom, monotony, errors, distress, illness, depression, worker turnover, absenteeism, and so on [21]. There are numerous practical solutions to cure sick jobs, including check-in work.

SOLUTIONS FOR MANAGEMENT, SOLUTIONS FOR WORKERS

The findings from the study of airport check-in workers can be applied to check-in systems of any design and level of mechanization in any country. The results justified making a series of recommendations, which I suggest should be discussed with workers and managers together. The results could be used to establish a benchmark for making improvements in airport check-in jobs.

The U.S. National Institute for Occupational Safety and Health identified major groups of risk factors for work-related MSDs [27]. Using these groups, a number of solutions are presented below for improving check-in workers' jobs.

Repetitiveness

Solutions:
- *Use mechanical aids* (such as installing roller bars at semi-mechanized systems to connect baggage weigh scales to conveyors to eliminate repeated lifting/carrying of bags)
- *Enlarge work content by adding more diverse activities* (such as safety professional on the ground duties, gate duties, providing assistance for passengers with special needs, possibly training to carry out other functions in airports, such as security checks)
- *Rotate workers* (such as ensuring workers do not carry out exclusively check-in duties for more than four hours per work shift, giving workers the possibility to organize their own shifts, ensuring that the same workers do not repeatedly work the shifts with the greatest number of flights)
- *Increase rest allowances* (such as a ten-minute break for every two hours of computer work, providing training so workers learn the benefits of alternating sitting and standing on the job, and encouraging postural change every two hours after a short break)

Force/Mechanical Stress

Solutions:
- *Decrease weight of tools/containers and parts* (such as 20 kg. universal maximum baggage limits)
- *Improve mechanical advantage* (such as zigzag surface on all mechanized conveyors and addition of barriers to prevent bags from falling off)
- *Use pads and cushions* (such as for standing work at the check-in counter)

Posture

Solutions:
- *Locate work to reduce awkward postures* (such as locating the baggage scale to prevent twisting in order to view it)
- *Alter position of tool* (such as computer keyboard) to avoid bending of wrist
- *Provide adjustable tables and chairs designed so that work can alternate between sitting and standing and so that furniture can be adjusted to suit the worker*
- Provide training on how to properly adjust work station furniture for individual size

Psycho-Social Stress

Solutions:
- *Enlarge workers' task duties* (such as combining check-in work with duties of safety professionals on the ground, duties at departure gates, client service, and services for passengers with special needs)
- *Allow more worker control over pattern of work* (such as allowing workers to establish schedules together, rather than imposing them)
- *Provide micro work pauses* (such as giving regular breaks of ten minutes every two hours)
- *Allow workers to participate in work station and task design, as well as work organization planning*

Violence

Solutions:
- *Establish a high check-in counter as the design standard to protect check-in workers against aggressive passengers and add counter height to protect workers where it does not exist*
- *Provide mandatory training for both workers and managers on dealing with aggressive passengers*
- *Establish a hierarchical structure at airports to enable and empower check-in workers to use their position as a first line of defense for passenger, crew, and aircraft safety.*

The findings from this study indicate that simple yet obvious preventive actions could be taken by management in any airport to protect airport check-in workers' health and well-being. The findings on MSDs and the levels of pain and suffering experienced by check-in workers should concern managers. The findings support the recommendation to install a fully mechanized conveyor system, or a roller bar at the very minimum, to eliminate the need for workers to lift and carry bags.

There are, of course, costs involved with installing a fully mechanized conveyor system. Information on the cost of installing and maintaining such a system was difficult to obtain in a precisely comparable manner for the three study sites. At Airport C in Switzerland, the installation cost was estimated at between 28 million and 31 million Swiss Francs (US$15-19 million) for the entire International Airport, which comprises 94 check-in work stations. The maintenance cost is estimated at between 1-1.5 million Swiss Francs per year (approximately US$1.8-2.7 million), which is approximately 2-3% of the purchase price of the mechanized baggage system. The costs included the entire system, from check-in to container loading for the airplanes, and included the costs of the mechanics, electronics, software, hardware, labor costs, and installation. The costs also included a machine that sorts bags by flight number. This device may be more sophisticated than what is needed at small airports. A smaller airport may be able to manage effectively with a simpler system, costing less money. In Québec and Ontario, Canada (Airports A and B), the cost of re-designing one fully mechanized work station was estimated at between C$25,000 and C$35,000 (US$16,250-22,750).

One key factor making the cost of the mechanized system high at Airport C was the cost of labor in Switzerland, which is higher than the cost of labor in most other countries in the world. It is therefore reasonable to assume that the re-design or installation and maintenance of a fully mechanized system would cost less in countries where the cost of labor is lower than in Switzerland.

Where costs prohibit immediate introduction of a fully mechanized system, preventive measures could begin with simple work station and work organization adjustments, often with no cost consequence to employers. Adding a roller bar to connect the check-in scale with the conveyor would be an important first step where the check-in area is not mechanized, to eliminate the need for workers to lift and carry every bag checked-in. Introducing wrist supports to the computer work station is a low-cost, simple solution to prevent the wrist from resting on the hard table surface. And regular rest breaks away from computer work should be introduced and respected.

Use of internationally recognized criteria for selecting or designing computer equipment and systems for check-in work also is suggested, as for computers used in any other workplace. Design and performance standards, such as those established by the International Organization for Standardization (ISO), provide managers with confidence that systems and equipment acquired can be used productively, safely, efficiently, and comfortably, and give workers a benchmark for judging their work conditions [83].

Findings from the study of check-in workers revealed that foot pain was experienced less often than other types of MSDs, but when foot pain was reported it was more severe among workers at Airport A in Ontario, where workers had to stand much more than did their counterparts at Airports B and C in Québec and Switzerland respectively. The foot and back pain reported by check-in workers provided further evidence to support the recommendation that workers be provided with sit/stand options during a work shift. Where chairs provided by employers were uncomfortable, not sturdy, or where there was a lack of leg room when sitting, workers were more likely to stand, which leads to such localized pain. Adding a cushioned mat on the standing surface is a simple and inexpensive means of giving relief to workers whose job entails any amount of standing. Check-in workers should be informed about the importance of good quality footwear, which supports the back, feet, and legs. The issue could be addressed by a health and safety committee, as well as when workers are informed about regulations on uniforms. A shoe allowance should be addressed where relevant.

Environmental factors that create discomfort for check-in workers should be addressed as well. Glare on check-in workers' computer screens can be eliminated, or at least reduced, by anti-glare screens that simply attach to the front of computer monitors, as well as through the use of adjustable monitors and tables enabling workers to adapt the angle of their computer. Double door systems in airports, or repositioning check-in work stations, may eliminate or at least ameliorate the problem. Management should recognize the aggravating effects on MSDs caused by extreme temperature fluctuations and make efforts to minimize them. Once noise as a design limitation is communicated to the engineering companies that develop and produce baggage conveyors, new generation conveyors could be engineered to use silent motors that reduce noise in the check-in work environment. In the meantime, airport management authorities and health and safety committee representatives could investigate with those same engineering companies whether a muffler could be used to reduce constant background conveyor motor noise.

The benefits of worker participation in the design of work stations, work processes, work organization, and in identifying and solving problems are cost effective in the long term. A first step would be for management to establish and promote a joint worker-management health and safety committee in every airport, to ensure that regular meetings of the committee are held, and that identified problems are addressed rapidly. The manner in which problems are addressed should be communicated to all workers. This approach is supported by research which showed that designing and implementing recommended changes jointly with management and workers prevents MSDs caused by prolonged computer work [50, 84, 85], and by strong reasons to suspect that lack of worker voice contributes to MSDs [47].

The effectiveness of joint health and safety committees should be examined at all airports. Obstacles should be addressed if committees do not function effectively, if they do not deal with and resolve priority issues, or if committee members do not contact the workers they represent to solicit information on a regular basis.

Additional benefits of worker participation can be seen when developing strategies that include coping with work-related violence [86, 87]. Joint worker-management strategies have been shown to protect the health and well-being of computer users, reduce costs to management from lost work time and sickness absence, as well as increase efficiency and productivity. Similar strategies can be extended to airport check-in workers, given the commonalities between their jobs and those of computer workers. "Workplace-people interaction is a continuous management responsibility and must be treated as such" [31, p. 229].

Improved communication, dialogue, and information exchange are badly needed between management and airport check-in workers, which was revealed by the difference between self-reported MSDs and officially recorded work-related injuries at all of the study sites and through interviews with workers. To be effective though, communication channels must exist for upward as well as downward communication. The direct participation of women, both collectively and individually, is especially necessary to report and interpret their experiences of work and health and to develop solutions to their work-related concerns.

Of no less importance is the need for management to pay attention to feedback from workers. A beneficial problem-solving process could be initiated through management discussing with workers their views on problems in job design. Airport check-in work is the self-expressed job of choice for many workers. This preference suggests that, in general, airport check-in workers have positive feelings about their jobs. A positive attitude toward one's job provides a good basis for opening channels of communication between management and workers and for involving workers in decision-making.

A lesson learned by management in one company after a strike in their main manufacturing plant was that a participative management style was not enough: ". . . we would have to learn to communicate better, because as much as anything, people's perceptions generate strikes" [1, p. 105]. If airport check-in workers go on strike, it represents a disaster for airlines. Flights would be cancelled, the public would be upset with the airlines for disruption to the services for which they have paid, customer loyalty to individual airlines may be damaged, travel plans would be upset, important meetings missed, vacations ruined, and so forth. The public blames the airlines for such happenings, not check-in workers or any other group of ground staff.

All working people need leverage to protect themselves from potential exploitation, from exposure to unacceptable conditions, to protest destructive management practices or abusive managers. Unionization is the key means of gaining such leverage. Where check-in workers are not unionized, it is the job of trade unions to organize such workers. In a democracy, legislation has to come from the grassroots, from the people the legislation is designed to protect. If working people do not cry out for help when they need it, help seldom is given voluntarily. More than ever, during this era where the focus is on individualism and voluntary initiatives by employers (versus statutory regulation), the importance of simultaneous top-down and bottom-up approaches cannot be overemphasized. Effective trade union action may be the only means to get recognition of MSDs and compensation for occupational diseases where they are not already compensated. The evidence on MSDs induced by check-in work conditions could be used as the basis for class action lawsuits, for example. Empirical evidence is food for many lawyers today. Clearly, however, this approach is not the desired course of action but should be seen as a last resort. Effective communication with employers and prevention with the full involvement of workers is much more effective in the long term.

How can managers know if they are truly involving workers in organizational decision-making? How can management know if it is enabling increased control and decision-making latitude for workers, with the goal of improved health outcomes, particularly for stress reduction? The following points can be used as a self-evaluation checklist [44, pp. 16-17], which can be applied to airport check-in workers:

- Work tasks: can the individual determine the order in which tasks are completed as well as choose among available methods for accomplishing them? Can the individual choose among a variety of tasks or projects? Can the individual influence the quality or quantity of his/her output?
- Work pacing: can the individual determine the pace of work or is it controlled by machine or management? Can the individual determine the scheduling and duration of rest breaks?
- Work scheduling: can the individual determine his/her work hours so that non-work demands can be better coordinated with work demands? Can vacations and days off be taken when the individual desires? Does the worker have a choice of shift schedules?
- Physical environment: is the individual able to modify, adjust, decorate, or otherwise personalize the work station? Does the individual have some control over lighting, temperature, noise, or privacy?
- Decision-making: does the individual have influence concerning organizational policies, goals, procedures, and choice of work methods, tools, and practices?
- Interaction: can the individual control the amount and timing of contact with other people, such as co-workers, customers, clients, or passengers?

- Mobility: does the individual perceive that he or she can leave the occupation or employer or does the individual feel locked in to a particular job?
- Skills development: does the individual have opportunities to develop new skills for continuous learning, to enlarge their job and the perspective with which they work?

The psycho-social work environment has been linked to the occurrence of MSDs, and appear to be more important for MSDs of the neck and shoulder region than the hand and wrist area [88]. A lack of management attention to workplace psycho-social factors and MSDs combined with a lack of worker voice and involvement in decision-making is not a recipe for improved worker health and productivity. Workers, like anyone else, appear to tense up their neck and shoulders when faced with emotional demands, psychological harassment, or job strain.

Based on the study's findings, the following simple recommendations are suggested to improve airport check-in workers' work conditions:

- Provide on-the-job training for computer users;
- Give training on proper lifting technique;
- Supply adjustable work station furniture;
- Offer training on how to adjust work station furniture where it is adjustable and on proper heights according to individual height and size;
- Design work scheduling for job rotation or job variation to provide time away from computer work;
- Provide regular rest periods to prevent continuous computer work throughout the work shift—approximately ten to fifteen minutes of rest is recommended for every hour or two of continuous computer work. Rest breaks should be taken away from the computer. With job variation, it can be accomplished by workers performing other tasks every couple of hours;
- Allow workers to choose when to take a break rather than having fixed rest schedules;
- Advise check-in workers to change positions, stand up, stretch, whenever they begin to feel tired, and to rest their eyes occasionally throughout the work shift;
- Provide chairs so that work can be performed alternately sitting or standing;
- Supply work stations that prevent wrists from resting on the edge of the keyboard, on a hard surface, or from bending at the wrist.

Airport B in Québec was the one study site where check-in workers had the possibility of performing their jobs alternating between sitting and standing, and workers made a nearly even break between the two postures. The fact that workers alternate sitting and standing over time may show the natural physical need for humans to adjust their body position on the job, provided the work station design allows this flexibility. Regular position change also can be an indication of back pain, which is often more painful when sitting. People in pain generally change their position frequently, trying to find a more comfortable position. At Airport A in Ontario, check-in work stations had no design flexibility and workers stood up throughout their entire work shift. Higher body load and MSDs of the neck, shoulder, and lower back were found at Airport A.

Management practices need to be examined with regard to their impact on airport check-in workers. The "quick turnaround" policies practiced by airlines increase physical strain, emotional demand, pressure, and workload for workers, without an accompanying increase in worker control over the organization of work or over any other aspect of the check-in job. The lack of worker control over the organization of their work was highlighted by two workers at the airport in Switzerland, who commented on the difficulty of irregular work schedules:

> The irregular schedules are a problem. We have shifts from 5 am to noon, and these change daily. It is difficult to live with, but it's part of the job. (Worker, Airport C).

> The irregular work schedules are a problem. (Worker, Airport C).

What appeared to be a complete lack of training and skills development available to airport check-in workers may also contribute to high rates of MSDs and worker turnover (such as at Airport C, where the average time spent on the job was half that of Airport A and B workers). The underutilization of skills has been negatively associated with workers' ability to adjust to the effects of high job demands, where worker control and decision latitude are low. The evidence pointed to implications for the quality of work performance, with the underutilization of skills as costly for employees, organizations, and companies [34, p. 232].

Work-related violence is a human issue that requires Human Resource Departments to play a leading role in prevention, intervention, and rehabilitation. The level of violence from aggressive passengers revealed in the study of check-in workers was a call for action. Increasingly, businesses are recognizing the need to invest in preventive programs to deter workplace violence. Employers need to address the problem of work-related violence against airport check-in workers as a service-sector job at the three levels described. To protect check-in workers, cabin crew, and passengers, a coordinated approach is needed.

To protect workers from aggressive passengers, an industry-wide approach involving governments, regulators, airlines, airports, and ground handling agents is needed. The ITF has issued guidelines recommending actions focused on:

- Preventive policies for airports, ground agents, and airlines aimed at minimizing the risks of disruptive passengers boarding an aircraft;
- Strategies and training for both crews and ground staff to manage incidents;
- Passenger information and service improvements;
- Sanctions against offenders;
- Post-incident support for staff and passengers who are victims of aggression;
- Changes in law to ensure law enforcement jurisdiction extends to all aircraft;
- Stronger laws covering these offences;
- An international treaty covering offenses in the air.

Further suggestions for protections against violence at work in general include trade union policy statements condemning verbal, physical, and physiological violence at work by employees or supervisors. Harassment clauses are recommended in collective bargaining language and in contract provisions covering worker health and safety. Legalistic solutions should be combined with efforts to eliminate the basic cause of the problem for women workers in particular, which is the inequality between women and men in the work force. Without power differences in the form of labor force sex segregation, unequal pay, and limited opportunities for the advancement of women, the possibility of all forms of violence would be drastically reduced [89].

Organizational commitment to the problem is essential. The necessary components of a company approach aimed at zero-level work-related violence include policies and procedures, employer commitment to the problem, security, crisis management teams, and employee assistance programs, as well as educating workers and managers about work-related violence [90]. Employers of airport check-in workers would be pro-active by providing regular and updated training on prevention of violence from aggressive passengers. The training should consist of:

- Developing interpersonal and communication skills for diffusing and preventing potentially threatening situations;
- Developing competence in the particular function to be performed;
- Improving the ability to identify potentially violent situations and people;
- Preparing a core group of mature and specially competent staff who can take responsibility for more complicated interactions;
- Developing violence-prevention strategies that include training managers and workers on how to recognize and handle violent behaviors and threats and aggressive words, tones, or behaviors. Strategies should include the development of mutual support among staff members and assistance from supervisors and co-workers [86, pp. 113-114].

The 2½-day training seminar on dealing with aggressive passengers that was initiated by the two employers of check-in workers at Airport C in Switzerland was an example of how employers of check-in workers can begin to tackle the problem of passenger violence. The seminar was empowerment-based and gave workers and managers specific tools for dealing with aggressive passengers. Participation by both workers and managers was mandatory. Information about long-term results or benefits associated with the training has not been available.

While this type of seminar can be a useful component in a violence-prevention program, alone it does not suffice. In the immediate future, airport employers could pay close attention to the physical environment and design of check-in work stations as a measure for protecting workers against violent passengers. The study of check-in workers showed that the high counter design of check-in work stations at Airport C in Switzerland served as a protective barrier against aggressive passengers. Installing entirely new, higher counters may be costly to management. A lower cost solution could be simply adding to the height of the existing counter with a heavy block, cut to size, or made of a lightweight composite material.

If airport check-in workers, and the travelling public, are to be protected against aggressive passengers, then employer action should aim at creating an environment where workers feel empowered, not powerless, have support systems in place, and are treated fairly. To develop preventive strategies against workplace violence, work organization and job content are additional key factors. Some of the more effective means to prevent aggression between workers and the public include appropriate staffing levels, task assignment based on experience and competence, clearly defined tasks, and work hours and shifts that are not excessive. Furthermore, it is important to modify management practices to limit customer dissatisfaction since it is strongly suspected to contribute in a major way to passenger aggressiveness. These factors should not be ignored in any management or union-led strategy to protect check-in workers from passenger violence.

Today, it is crucial that workers responsible for public safety be specially trained, particularly in airports and other public transport facilities. The events of September 11, 2001 in the United States highlighted a new era, where adequate training is crucial, putting into question the types of training workers received in the past. In this new era, inadequate training or worker performance can lead to disasters. The problem of aggressive passengers is significant and increasing, requiring action in airports all over the world to increase protection for workers and passengers and to diminish negative effects for employers. The prevention of air rage and general safety in the air can easily begin with check-in workers, the natural safety professionals on the ground. Airport check-in workers are a logical first line of defense for public safety in airports. Given the necessary training to empower them to take on this critical role with management support, check-in workers are well placed to identify potentially aggressive or threatening passengers, and to set off a chain of actions aimed at preventing aggression toward airport security staff or disasters in the air. This function should be critical in an enlarged job scope that encompasses being a safety professional on the ground.

Adoption of a universal policy on maximum allowable weight of checked baggage may be a useful step. North America and South America allow baggage to be checked up to 33 kilos per bag for all passengers (maximum 2 checked bags, totaling 66 kilos), while many other countries limit checked baggage to 20 kilos total per passenger in economy class. (Economy class passengers make up the majority of travelers in all countries.) The 33-kilos-per-bag limit means a higher load on the musculoskeletal system of check-in workers. Stricter limitations on baggage weight are starting to be imposed in various countries.

A flexible standards-based approach to occupational health and safety in airport check-in work also may be worth consideration. Conditions involving awkward movements, complex tasks, and physical strain do not necessarily lend themselves to a rigid standard-setting process. More flexible approaches are required to ensure that workers' health is protected and to prevent employers from bearing the financial burden related to poor or ill-adapted work conditions.

An example of a flexible and innovative standards-based approach was proposed through the Workers' Compensation Board of British Columbia, Canada. The board proposed a forward-looking draft ergonomics regulation that included no specific standards; rather its focus was for employers to identify factors that might expose workers to a risk of adverse health effects. The employer would then have been required to eliminate or minimize the risk of adverse health effects to workers [61]. This kind of approach could be used in place of fixed weight limits for lifting, for example, which may not be protective for workers of all sizes and strength levels.

A similar approach could be particularly useful in airports in developing countries where national health and safety legislation may not exist, and where awareness of ergonomics may be limited or non-existent. A checklist of potential risk factors in the hands of both managers and workers could be useful where legislation does not exist or where it is not protective of check-in workers. Such an approach is suggested as a means of establishing flexible regulations, which could be applied to the needs of check-in work, supermarket check-out work, and computer clerical work.

An additional suggestion for future trade union action would be to consider developing and carrying forward skills objectives. These can range from individual technical skills, such as how to lift properly, to group action skills, such as how to advocate for re-design of the workplace. Skills objectives could be useful in analyzing check-in jobs in any airport, and in proposing ergonomic modifications where social action is needed to initiate and implement changes in work organization through labor-management cooperation.

The study team found that MSDs among check-in workers increase with time on the job: the more years one works at the check-in, the more MSD symptoms appear. Such findings are not surprising given the evidence showing that MSDs are cumulative over time. It may be tempting for management to use such findings to justify using low wages and poor work conditions, to encourage check-in worker turnover after an identified period of time on the job. This would constitute a short-term investment vision, which is not cost-effective over time. Workplace-based health registries are a way to quickly accumulate evidence to demonstrate to management that a short-term vision of investment in human resources is not necessarily cost effective. Improving work conditions is more cost-beneficial over time than not paying attention to work conditions and is a more human approach to work.

The findings from this study support the need to look at the entire concept and all job tasks involved in airport check-in work from a holistic perspective to ensure that workers can perform a check-in job safely for as long as they want. With skills development, training and recognition of the hazards inherent in the check-in job and preventive measures can be built in to protect workers' health.

Airport check-in workers, their respective trade unions, and management could explore job re-design, including a proper "balancing" of the organization of work. Involving workers in organizational decision-making, such as through their involvement in work station design improvements, may enhance positive aspects of the job, leading to a reduction in physical and emotional stress and improved health outcomes.

A final suggestion addresses the civil aviation industry as a whole. Airlines could consider increasing the cost of airplane tickets and use the new revenue for training and preventive actions. However, market competition makes it unlikely to happen at present.

Ultimately, for change to take place, some degree of "buy in" or "ownership" from those with decision-making power and from those whom change is meant to impact is needed. Conducting action-oriented research together with organized bodies, such as national level or international level trade unions, which communicate with and involve workers, their representatives, and management at the local level, is both useful and equitable as an approach. With engagement from these various groups, the potential for change is greatly increased.

FOOD FOR THOUGHT . . .

Modern management practices and technology are practical only with the increase in production. This is a vicious cycle—continuously increasing production through the application of technology to speed up work processes. Management practice both facilitates and necessitates the reduction of labor in the same production processes. This approach may drive capitalism, but the same cycle generates negative effects on workers' health through the speed-up and the constant insecurity. In addition, there are multiple adverse consequences for societal well-being. Consider that simple cost-benefit formulas taking into account the often hidden aspects of ill health and psycho-social suffering have not been developed for the public domain. This fact can only be explained by socio-political forces because the results of such formulas would surely imply that the time has come for a major change in patterns of production and consumption, just when both are at their historical highs globally, with new markets being captured all the time. Most companies perform some degree of cost-benefit analyses for expenses incurred, whether crude or sophisticated. But few conduct even simple research at a workplace level to identify what are often the hidden costs related to work conditions and management practices. Here is where one can find the surprising stories of what workers live with daily, and what often explains absenteeism, turnover, decreased productivity, and reduced morale.

The rationale for Taylor's management practices was to force workers to do all the work they could without injury to their health at a pace that could be sustained throughout a working lifetime. Today, few longitudinal studies exist to prove whether the amount and pace imposed by a particular job causes no injury to a worker's health over a lifetime. Today such studies are difficult to design since few workers perform only one job for their entire work life.

Notwithstanding, we have sufficient evidence to indicate that the workloads and ever-increasing pace of work do, in fact, cause injury to workers. It is, therefore, time to not only reduce load and pace, but to re-think work processes and the organization of work, which cannot be done by management without the full involvement of workers.

ENDNOTES

1. R. Semler, *Maverick: The Success Story Behind the World's Most Unusual Workplace*, Warner Books, New York, p. 98, 1993.
2. P. Landsbergis, P. Schnall, and J. Cahill, The Impact of Lean Production and Related New Systems of Work Organization on Worker Health, *Journal of Occupational Health Psychology, 4*, pp. 108-130, 1999.
3. E. Skorstad, Lean Production, Conditions of Work and Worker Commitment, *Economic and Industrial Democracy, 15*, pp. 429-455, 1994.
4. J. R. Hackman, G. Oldham, R. Janson, and K. Purdy, A New Strategy for Job Enrichment, *California Management Review, 17*:4, pp. 57-71, 1975.
5. Canadian Autoworkers' Union, *CAW Automotive Sector Study: Benchmarking Working Conditions (Vehicle Assembly Plants)*, Ontario, Canada, May 1996.
6. J. P. Womack, D. T. Jones, and D. Roos, *The Machine That Changed the World: The Story of Lean Production*, Rawson Associates, New York, p. 101, 1990.
7. C. Korunka, A. Weiss, K-H. Huemer, and B. Karetta, The Effect of New Technologies on Job Satisfaction and Psychosomatic Complaints, *Applied Psychology: An International Review, 44*:2, pp. 123-142, 1995.

8. P. Landsbergis, The Changing Organization of Work and the Safety and Health of Working People: A Commentary, *Journal of Occupational and Environmental Medicine, 45*:1, pp. 61-71, January 2003.
9. S. Kakar, *Frederick Taylor: A Study in Personality and Innovation*, Cambridge, Massachusetts, 1970.
10. H. Braverman, *Labor and Monopoly Capital: The Degradation of Work in the Twentieth Century*, Monthly Review Press, New York and London, 1974.
11. M. Grenier-Peze, Chroniques de la Violence Ordinaire: Organization Psychique Individuelle, Organization du Travail, Organization du Système de Soin, *Travailler, 4*, pp. 1-12, 2000.
12. J. F. Chanlat, Nouveaux Modes de Gestion, Stress Professionnel et Santé au Travail, in *L'homme à l'echine Plié*, I. Brunstein (dir), Desclee de Brouwer, pp. 29-60, 2000.
13. Advertising for anti-depressants and anti-anxiety medicines is widespread today, increasingly so in developing and transitional economies as well as in rich countries. Advertisements portray happy, smiling people, giving the impression that taking the medicine will make you feel happy too. Yet more and more people live with growing insecurities today (*Economic Security for a Better World*, ILO, Geneva, 2004) accompanied by the fears that coincide with increasing insecurity in one's life and reduced social support. The marketing and advertising of "feel-better" products serve to individualize problems that are often rooted in objective factors beyond the control of individuals. Working people may feel depressed because they are harassed or treated as an object at work. They may feel out of control because work is intensifying in speed, quantity and demand level, combined with a lack of employment security, and no control over conditions in the workplace. For many the solution is anti-depressant and/or anti-anxiety medication in order to cope with going to work every day. While the factors changing the work environment are objective, driving workers to increased levels of stress with resultant negative health outcomes, many feel that what is happening to them is their fault, that they did something wrong. Such internalization is self-destructive and counter-productive to collective change.
14. R. Karasek, D. Baker, F. Marxer, A. Ahlbom, and T. Theorell, Job Decision Latitude, Job Demands and Cardiovascular Disease: A Prospective Study of Swedish Men, *American Journal of Public Health, 71*:7, pp. 694-705, 1981.
15. A. Bergmann, Consequences Sociales et Humaines d'une Concurrence Effrénée, in *Revue Economique et Sociale*, pp. 211-224, December 1994.
16. A. Bergmann, The Manager as a Teacher: What He Knows and Should Know About Learning, in *Die Förderung der Leistungsbereitschaft des Mitarbeiters als Aufgabe der Unternehmensführung*, Ch. Lattmann, G. Probst and F. Tapernoux, Physica Verlag, Heidelberg, pp. 27-43, 1992.
17. M. Grenier-Peze, Contrainte par Corps: Le Harcèlement Moral, in *Travail, Genre et Sociétés. Dossier: Harcèlement et Violence, Les Maux du Travail:5, L'Harmattan, Paris*, pp. 1-9, 2001.
18. J-F.Chanlat, Stress, Psychopathologie du Travail et Gestion, in *L'individu dans L'organization: Les Dimensions Oubliées*, J-F. Chanlat (ed.), Presses de l'université Laval/Eska, Sainte Foy, Paris, pp. 709-721,1990.

19. Free translation from H. Mintzberg, *Mintzberg on Management: Inside Our Strange World of Organizations*, The Free Press, New York, 1989, in J. F. Chanlat, Nouveaux Modes de Gestion, Stress Professionnel et Santé au Travail, in *L'homme à L'echine Plié*, I. Brunstein (dir), Desclee de Brouwer, p. 53, 2000.

20. S. Hecker (ed.), Management and Policy: Introduction and Overview to Education and Training, in *Encyclopaedia of Occupational Health and Safety* (4th Edition), International Labor Office, Geneva, 1, Ch. 18, pp. 2-4, 1998.

21. S. Snook, The Practical Application of Ergonomics Principles, in *Minesafe International*, 1993.

22. National Academy of Sciences, *Musculoskeletal Disorders and the Workplace: Low Back and Upper Extremities*, Washington, D.C., 2000.

23. A. Kilbom, Intervention Programmes for Work-Related Neck and Upper Limb Disorders: Strategies and Evaluation. *Ergonomics, 38*, pp. 922-927, 1998.

24. EU Advisory Committee on Safety, Hygiene and Health at Work, May 2001.

25. R. Bergamasco, C. Girola, and D. Colombini, Guidelines for Designing Jobs Featuring Repetitive Tasks. *Ergonomics, 41*:9, pp. 1364-1383, 1998.

26. J. Benson, Control of Low Back Pain in Industry Through Ergonomic Redesign of Manual Materials Handling Tasks. in *Trends in Ergonomics/Human Factors III*, W. Karwowski (ed.), Elsevier, Amsterdam, 1986.

27. National Institute for Occupational Safety and Health, *Cumulative Trauma Disorders in the Workplace*, NIOSH, U.S. Department of Health and Human Services, NIOSH Publication No. 95-119, 1995.

28. J. V. Johnson, Occupational Stress, in *Preventing Occupational Disease and Injury* (2nd Edition), J. L. Weeks, B. S. Levy, and G. R. Wagner (eds.), The American Public Health Association, Washington, D.C., 2004.

29. M. Kompier, The Psychosocial Work Environment and Health—What Do We Know and Where Should We Go? *Scandinavian Journal of Work, Environment, and Health, Editorial, 28*:1, pp. 1-4, 2002.

30. J. H. Browne, Benchmarking HRM Practices in Healthy Work Organizations, *American Business Review, 18*:2, pp. 50-61, 2000.

31. D. Carnevale and J. M. Rios, How Employees Assess the Quality of Physical Work Settings, *Public Productivity and Management Review, 18*:3, p. 229, 1995.

32. A. Bergmann, Worker Participation, in *Levels of Analysis for Human Behavior in Organizations*, p. 392, 1978.

33. J. V. Johnson, Empowerment in Future Worklife, *Scandinavian Journal of Work, Environment and Health, 23*:4, pp. 23-28, 1997.

34. L. D. Sargent and D. J. Terry, The Effects of Work Control and Job Demands on Employee Adjustment and Work Performance, *Journal of Occupational and Organizational Psychology, 71*, pp. 219-236, 1998.

35. A. Aittomaki, E. Lahelma, and E. Roos, Work Conditions and Socio-Economic Inequalities in Work Ability, *Scandinavian Journal of Work, Environmental and Health, 29*:2, pp. 159-165, 2003.

36. *Economic Security for a Better World*, International Labour Office, Geneva, 2004.

37. J. Siegrist and M. Marmot, Health Inequalities and the Psychosocial Environment—Two Scientific Challenges, in *Social Science and Medicine, 58*:8, pp. 1463-1473, 2004.

38. B. G. Link and J. Phelan, Social Conditions as a Fundamental Cause of Disease, *Journal of Health and Social Behavior*, Extra Issue, pp. 80-94, 1995.

39. For more discussion, see my analysis of the performance of 95 countries (governments) on the protection of workers' health and safety presented as chapter 7 of *Economic Security for a Better World*, International Labour Office, Geneva, 2004. The analysis demonstrates that countries do not appear to achieve even minimally acceptable levels of protection of workers' health and safety in the absence of basic economic security measured as income security plus voice representation security.

40. H. S. Shannon, J. Mayr, and T. Haines, Overview of the Relationship Between Organizational and Workplace Factors and Injury Rates, *Safety Science, 26*:3, pp. 201-217, 1997.

41. A. Imada, Macroergonomic Approaches for Improving Safety and Health in Flexible, Self-Organizing Systems, in *The Ergonomics of Manual Work: Manual Handling and Information Processing at Work*, Taylor and Francis, London, pp. 477-480, 1993.

42. P. Jensen, Can Participatory Ergonomics Become the Way We Do Things in this Firm—The Scandinavian Approach to Participatory Ergonomics, *Ergonomics, 40*:10, pp. 1078-1087, 1997.

43. R. Karasek and T. Theorell, *Healthy Work: Stress, Productivity and the Reconstruction of Working Life*, Basic Books, New York, 1991.

44. D. Ganster, Worker Control and Well-Being: A Review of Research in the Workplace, in *Job Control and Worker Health*, S. L. Sauter, J. J. Hurrell, and C. Cooper (eds.), John Wiley and Sons, Chichester, pp. 3-23, 1989.

45. R. Karasek, Control in the Workplace and Its Health-Related Aspects, in *Job Control and Worker Health*, S. L Sauter, J. J. Hurrell, C. Cooper (eds.), John Wiley and Sons, Chichester, pp. 129-159, 1989.

46. C. Dejours, Nouveau Regard sur la Souffrance Humaine dans les Organizations, in *L'individu dans L'organization: Les Dimensions Oubliées*, J-F. Chanlat (ed.), Presses de l'université Laval/Eska, Sainte Foy, Paris, pp. 687-708, 1990.

47. J. Faucett and D. Rempel, VDT-Related Musculoskeletal Symptoms: Interactions Between Work Posture and Psychosocial Work Factors, in *American Journal of Industrial Medicine, 26*, pp. 597-612, 1994.

48. R. Karasek, Job Demands, Job Decision Latitude, and Mental Strain: Implications for Job Redesign, *Administrative Science Quarterly, 24*, pp. 285-308, 1979.

49. L. Alfredsson, C. Spetz, and T. Theorell, Type of Occupation and Near Future Hospitalization for Myocardial Infarction and Some Other Diagnoses, *International Journal of Epidemiology, 14*:3, pp. 378-388, 1985.

50. A. Statham and E. Bravo, The Introduction of New Technology: Health Implications for Workers, *Women and Health, 16*:2, pp. 105-129, 1990.

51. K. Grant and D. Habes, An Analysis of Scanning Postures Among Grocery Cashiers and Its Relationship to Checkstand Design, *Ergonomics, 38*:10, pp. 2078-2090, 1995.

52. I. Niedhammer, M. F. Landre, A. LeClerc, F. Bourgeois, P. Franchi, J. F. Chastang, G. Marignac, P. Mereau, D. Quinton, C. R. Du Noyer, A. Schmaus, and C. Vallayer, Shoulder Disorders Related to Work Organization and Other Occupational Factors Among Supermarket Cashiers, *International Journal of Occupational and Environmental Health, 4*:3, pp. 168-178, 1998.

53. S. Toon, Service Industries Say Safety Plays Key Role in Putting Customer First, *Occupational Health and Safety, 62*:11, pp. 61-69, 1993.

54. M. Vézina, L. Geoffrion, C. Chatigny, and K. Messing, A Manual Materials Handling Job: Symptoms and Working Conditions Among Supermarket Cashiers, *Chronic Diseases in Canada, 15*:1, pp. 17-22, 1994.

55. K. Fredriksson, L. Alfredsson, C. Bildt Thorbjornsson, L. Punnett, A. Toomingas, M. Torgen, and A. Kilbom, Risk Factors for Neck and Shoulder Disorders: A Nested Case-Control Study Covering a 24-Year Period, *American Journal of Industrial Medicine, 38,* pp. 516-528, 2000.

56. C. Korunka, A. Weiss, K-H. Huemer, and B. Karetta, The Effect of New Tech-nologies on Job Satisfaction and Psychosomatic Complaints, *Applied Psychology: An International Review, 44*:2, pp. 123-142, 1995.

57. G. S. Lowe, G. Schellenberg, and H. S. Shannon, Correlates of Employees' Per-ceptions of a Healthy Work Environment, *American Journal of Health Promotion, 17*:6, pp. 390-399, 2003.

58. R. M. Sapolsky, *Why Zebras Don't Get Ulcers: An Updated Guide to Stress, Stress-Related Diseases, and Coping,* W. H. Freeman and Company, New York, pp. 265-266, 1994 and 1998.

59. M. J. Smith and P. C. Sainfort, A Balance Theory of Job Design for Stress Reduction, *International Journal of Industrial Ergonomics, 4,* pp. 67-79, 1989.

60. B. Gardell, Alienation and Mental Health in the Modern Industrial Environment, in *Society, Stress, and Disease,* L. Levi (ed.), Oxford University Press, London, pp. 148-180, 1971.

61. See for example, M. Smith, 1997; P. Sainfort, Job Design Predictors of Stress in Automated Offices, in *Behaviour and Information Technology, 9*:1, pp. 3-16, 1990; R. Karasek and T. Theorell, 1991; S. Sauter, M. Dainoff, and M. Smith (eds.), *Promoting Health and Productivity in the Computerized Office,* Taylor and Francis, London, 1990; J-F.Chanlat, 1990; S. Deutsch, 1989; R. Baker and N. Wallerstein, Management and Policy: Worker Education and Training, in *Encyclopaedia of Occu-pational Health and Safety* (4th Edition), International Labor Office, Geneva, 1, Ch. 18, pp. 7-12, 1998; K. Messing (ed.), *Integrating Gender in Ergonomic Analysis: Strategies for Transforming Women's Work, European Trade Union Technical Bureau for Health and Safety,* Brussels, 1999; A. Bergmann, 1994; B. Silverstein, E. Welp, N. Nelson, and J. Kalat, Claims Incidence of Work-Related Disorders of the Upper Extremities: Washington State, 1987 through 1995, in *American Journal of Public Health, 88*:12, pp. 1827-1833, 1998.

62. S. Deutsch, Worker Learning in the Context of Changing Technology and Work Environment, in *Learning at Work: A New Approach to the Learning Process in the Workplace and Society,* H. Leymann and H. Kornbluh (eds.), Gower, 1989.

63. J. R. Hackman, G. Oldham, R. Janson, and K. Purdy, A New Strategy for Job Enrichment, in *California Management Review, 17*:4, pp. 57-71, 1975.

64. K. Kogi and R. N. Sen, Third World Ergonomics, *International Reviews of Ergonomics, 1,* pp. 77-118, 1987.

65. A. Hochschild, *The Managed Heart: Commercialization of Human Feeling,* Univer-sity of California Press, San Francisco, 1983.

66. R. Firth, *The Primitive Polynesian Economy*, Routledge, London, 1939.

67. P. Freire, *Pedagogy of the Oppressed*, Continuum Press, New York, 1970.
68. P. Freire, *Education for Critical Consciousness*, Continuum Press, New York, 1973.
69. See for example, R. Baker and N. Wallerstein, Management and Policy: Worker Education and Training, in *Encyclopaedia of Occupational Health and Safety* (4th Edition), International Labor Office, Geneva, 1, Ch. 18, pp. 7-12, 1998; K. Messing, *One-Eyed Science: Occupational Health and Women Workers*, Temple University Press, Philadelphia, 1998; E. Rosskam, Women Moving Mountains: Women Workers in Occupational Safety and Health, in *Women and Environment International Journal, 48/49*, University of Toronto, 2000; N. Wallerstein and M. Weinger, Health and Safety Education for Worker Empowerment, *American Journal of Industrial Medicine, 22*, pp. 619-635, 1992; K. De Koning and M. Martin (eds.), *Participatory Research in Health: Issues and Experiences*, Zed Books, London, 1996; H. Abrams, The Worker as Teacher in Commentary, *American Journal of Industrial Medicine, 4*, pp. 759-768, 1983; S. Snook, 1993; S. Hecker (ed.), 1998; K. Kogi, and R. N. Sen, 1987; S. Deutsch, 1989; A. Bergmann, 1992.
70. M. Smith, Psychosocial Aspects of Working with Video Display Terminals (VDTs) and Employee Physical and Mental Health, *Ergonomics, 40*:10, pp. 1002-1015, 1997.
71. National Institute for Occupational Safety and Health, *Participatory Ergonomic Interventions in Meatpacking Plants*, NIOSH, U.S. Department of Health and Human Services, NIOSH Publication No. 94-124, 1994.
72. National Academy of Sciences, *Work-Related Musculoskeletal Disorders: A Review of the Evidence*, Washington, D.C., 1998.
73. P. C. Bohr, B. A. Evanoff, and L. D. Wolf, Implementing Participatory Ergonomics Teams Among Health Care Workers, *American Journal of Industrial Medicine, 32*, pp. 190-196, 1997.
74. M. Nagamachi, Participatory Ergonomics: A Unique Technology of Ergonomics Science, in *The Ergonomics of Manual Work: Manual Handling and Information Processing at Work*, Taylor and Francis, London, pp. 41-48, 1993.
75. M. Frese, Theoretical Models of Control and Health, in *Job Control and Worker Health*, S. L. Sauter, J. J. Hurrell, and C. Cooper (eds.), John Wiley and Sons, Chichester, p. 113, 1989.
76. E. E. Adam Jr. and R. J. Ebert, *Production and Operations Management: Concepts, Models and Behavior* (5th Edition), Prentice Hall, New Jersey, p. 15, 1992.
77. B. Gustavsen and G. Hunnius, *New Patterns of Work Reform: The Case of Norway*, Universiteitsforlaget, Oslo, 1981.
78. Joint Industrial Safety Council, SAF, LO, PTK, *Working Environment Agreement*, Stockholm, 1984.
79. M. Satherley and N. Thomas, Successful Ergonomics—Implementing a Programme Within a Small Manufacturing Plant, in *The Ergonomics of Manual Work: Manual Handling and Information Processing at Work*, Taylor and Francis, London, pp. 517-518, 1993.
80. L. Patry, M. G. Costa, and I. Kourinka, Participatory Ergonomics and Prevention of Low Back Pain in Manual Material Handling, in *The Ergonomics of Manual Work: Manual Handling and Information Processing at Work*, Taylor and Francis, London, pp. 523-526, 1993.

81. J. Forester, *Planning in the Face of Power*, University of California Press, Berkeley, 1989.
82. F. Steele, *Physical Settings and Organization Development,* Addison-Wesley, Reading, Massachusetts, p. 8, 1973.
83. T. F. M. Stewart (ed.), Ergonomics Standards, in *Encyclopaedia of Occupational Health and Safety* (4th Edition), International Labor Office, Geneva, 2, Ch. 52, pp. 31-34, 1998.
84. M. J. Smith, B. G. F. Cohen, and L. W. Stammerjohn Jr., An Investigation of Health Complaints and Job Stress in Video Display Operations, *Human Factors, 23*:4, pp. 387-400, 1981.
85. S. Smith, D. Willms, D. Johnson, (eds.), *Nurtured by Knowledge: Learning To Do Participatory Action-Research*, The Apex Press, New York, 1997.
86. D. Chappell and V. Di Martino, *Violence at Work*, International Labor Office, Geneva, 2000.
87. K. A. Rogers and D. Chappell, *Preventing and Responding to Violence at Work*, International Labor Office, Geneva, 2003.
88. T. R. Hales, S. L. Sauter, M. R. Peterson, L. J. Fine, V. Putz-Anderson, L. R. Schleifer, T. T. Ochs, and B. P. Bernard, Musculoskeletal Disorders Among Visual Display Terminal Users in a Telecommunication Company, *Ergonomics, 37*:10, pp. 1603-1621, 1994.
89. P. Crull, Sexual Harassment and Women's Work, in *Double Exposure: Women's Health Hazards on the Job and at Home*, W. Chavkin (ed.), Monthly Review Press, New York, 1984.
90. E. Chenier, The Workplace: A Battleground for Violence, *Public Personnel Management, 27*:4, pp. 557-568, 1998.

CHAPTER 7

Beyond The Previously Known

We are paid to do a job, not to have the job "do us in" [1].

A campaign has been underway for years to convince working people that they are involved; that their voice is heard in the workplace; that Taylorism is a relic of the past; that MSDs, psychological distress, and a range of health problems are not work-induced or work-related; and that women's health issues at work are recognized and important. This study has tried to cut through that campaign's crusty build-up from years of talk.

The airport check-in worker study investigated a largely female-dominated occupation, previously unstudied, and the findings revealed that many hidden job hazards affect workers' health and well-being in what appears on the surface to be a "safe" job. Some claim that work has changed from manual to mental labor due to the increased use of information and communications technology, the rapid expansion of services, the globalization of the economy, the changing structure of the work force, and the increased flexibility of work [2]. Yet investigations of many jobs perceived as primarily mental toil, assumed not to be manual and physical in nature, reveal the opposite. It is especially true about jobs performed by women, whose manual and physical work does not decrease as women workers age.

The approach used in this exploratory two-country study can be applied to conduct similar cost-effective research on other occupations for which occupational health research has not been undertaken. Many such occupations appear to be "safe" but actually have "hidden" hazards. This description is typical of jobs done by women and jobs in the service sector where the hazards are still not recognized due to a lack of investigation of women's work. Looking at jobs from a holistic perspective is essential to get a clear view of the management practices that have such dramatic effects on workers' health as well as the more obvious physical exposures.

Because airport check-in work was not previously studied, comparing the occupation with other jobs that have important common characteristics has

231

allowed me to make conclusions about the "hidden" hazards inherent in check-in work. Creating a comparative foundation is useful where no body of literature exists. This approach enabled me to demonstrate that the hidden hazards in check-in work commonly exist in high-demand and low-control service-sector jobs and among airport baggage handlers whose work involves physical lifting. Using what is known about three comparison groups has helped to fill in the knowledge gap about check-in workers.

Airport check-in work compares with the kind of work performed by supermarket check-out and computer clerical workers, both similar to check-in work in the high-demand/low-control aspects that typify these jobs, which most often are performed by women. They impose static postures on workers—either sitting or standing. All three jobs commonly are associated with poorly designed, ill-adapted work stations, work tasks, and systems of work organization. Both supermarket check-out work and computer clerical work are clearly associated with high rates of MSDs; computer clerical workers have been shown to have high rates of stress. All three jobs present additional risks and strain from dealing with the public. Supermarket check-out workers, computer clerical workers who work with the public, and airport check-in workers have to sell their "emotional labor" as part of their job. The similarities of these jobs leave little reason to doubt that they share similar causes of the MSDs, physical strain and psychological distress reported by check-in workers.

Similarities between airport check-in workers and airport baggage handlers also have been established. It is clear that the weights lifted by airport baggage handlers, a predominantly male group, and the attention given to their musculoskeletal health, particularly their backs, as well as high rates of lost work time due to back injuries, demonstrate the need for greater attention to check-in workers. Airport check-in workers perform many of the same types of lifting and carrying functions as baggage handlers. Prevention programs for airport baggage handlers should focus on low back pain and coronary artery disease [3]. Given the correspondence in job functions and exposures (lifting and carrying baggage, twisting and awkward postures), the same preventive recommendations should be extended to airport check-in workers.

The prevalence of MSDs reported by check-in workers provides a picture of injury, pain, and lost work time due to work-related MSDs. Employer-provided lost work time injury reports do not reflect the reality of workers' injuries and suffering. Other studies also have shown that a significant proportion of work-related MSDs are never reported as such, leading employers to believe that the rates of MSDs experienced by their employees are far lower than they are in reality [4]. The non-reporting and underreporting of injuries effectively hides the real costs of work-related MSDs. It seems reasonable to assume that some of check-in workers' officially registered injuries that result in lost work time are likely related to MSDs.

I figured that check-in workers in airports that have semi-mechanized systems for baggage check-in would experience significantly more negative health outcomes, particularly more MSDs, than check-in workers at airports with fully mechanized systems. The results have been revealing. A variety of factors and characteristics inherent in check-in work at any type of system are likely causes of MSDs and lost work time. High rates of MSDs and suffering at all types of check-in systems (mechanized and semi-mechanized) are due to lifting and carrying bags, static posture, twisting, and awkward postures from manipulating bags on the conveyor belt. A high risk of violence, both real and perceived, due to increases in aggressive passengers, increases in work intensity, lack of protective management policies, and quick turnaround policies create the same pressures of work rhythm everywhere, independent of the level of mechanization for checking in baggage. These findings suggest that the entirety of check-in work must be addressed, not simply the baggage check-in system.

While direct costs for lost work time and registered injuries can be measured, pain that interferes with sleep, that inhibits the ability to perform one's job at an optimum level (particularly in a friendly manner) and that is disruptive to activities in daily life outside of work, also bear a cost, albeit harder to measure. These indirect costs are borne mainly by workers and their families, but are a cost to management as well, given that productivity, alertness on the job, customer satisfaction, and efficiency all can be reduced when workers are sleep deprived or when muscular pain restricts their freedom of movement.

GENERAL CONCLUSIONS: VIOLENCE AND MSDs

Probably the two most important findings of this study are the existence and degree of work-related violence faced by check-in workers and the prevalence and severity of MSDs, neither of which previously has been described. During their employment as a check-in worker, this study found that one in twenty workers had been physically assaulted on the job by aggressive passengers and 80% had been exposed to verbal abuse from passengers. The level of perceived risk of violence is also substantial in this occupation; nearly half of the workers reported that they see violence as a substantial risk factor at their jobs, an underlying stressor that they go to work with every day.

The frequency of violence experienced by check-in workers in two different countries and at three different types of airports, is significant and consistent with the risk of violence in other service-sector jobs. This study considered violence from disruptive passengers in the form of verbal abuse, threats, and/or physical assault. Few check-in workers told how often they had had such experiences, which, understandably, may be a sensitive or even traumatic issue for individuals to detail.

These findings should ring alarm bells for management and suggest the need for empowerment-based training for check-in workers and managers on how to deal with aggressive or unruly passengers. These findings highlight as well the need for management systems to be put in place in every airport to protect workers, passengers, and crew from potentially violent passengers.

> Unless airlines and passenger handling agencies actively promote and reinforce the safety professional role of employees in the minds of passengers, such staff will less easily be able to enforce their authority or get passengers to comply with their instructions. The active promotion of staff as safety professionals is an essential precondition to their empowerment . . . [5].

MSDs are prevalent and severe among check-in workers and may lead to temporary or permanent disability. An important number of workers live with pain from MSDs in various parts of their bodies. Some workers have lost work time due to pain or disability. A substantial number of check-in workers experience pain that interferes with their job performance, while many perform their job functions despite living with significant pain. Check-in workers lose sleep due to musculoskeletal pain and MSDs cause significant disruption to their activities outside of work.

The increase in MSDs with years performing the job of check-in worker is not to suggest that the number of years employed at airport check-in work should be limited as a preventive measure. On the contrary, the study findings support the need to use a systems approach to examine the entire concept and job tasks involved in check-in work in order to ensure that workers can perform the job safely for many years. With training and recognition of the hazards inherent in the check-in job, preventive measures can be built in to protect workers' health. For example, findings revealed a significant number of workers who had an adjustable chair at the check-in yet still suffered from MDSs. This fact indicated that various causal factors appear to interact, resulting in MSD symptoms.

These findings showed that airport check-in work has clear hazards associated with the job, an occupation which is likely to cause severe MSDs. None of the workers in the study received training on proper lifting technique, even though lifting, carrying, or handling baggage appears to be sufficient to cause MSDs, even at fully mechanized check-in work stations. The awkward twisting and bending involved in baggage tagging also appears to cause MSDs. It is a cause for concern when 24 workers report living with pain in four different body points at once and when only some 20% of workers reported not having any musculoskeletal pain.

Job factors that may be linked to MSDs among airport check-in workers include objective physical factors as well as psycho-social factors, which can affect check-in workers anywhere. While the perception and effects of psycho-social factors may vary among different cultures, violence from aggressive passengers, for example, is a global phenomenon [5, p. 12]. The potential for MSDs and serious strain-related effects to occur among airport check-in workers exists in every country. Differing social and cultural factors notwithstanding, check-in workers in any country are at risk of developing MSDs when they lift and carry baggage during their entire work shift, where the work organization and work station design require a static seated or standing position during the entire work shift, where workers work on computers at non-adjustable work stations, or where awkward postures are adopted for tagging and moving bags along conveyor belts. Such conditions are similar to those experienced by the workers investigated at the three airports in Canada and Switzerland.

And check-in workers in any airport face the same risks of developing strain and distress-related ill health and MSDs from the lack of control over their jobs, the lack of participation in organizational decision-making, quick turn-around policies, increasing pressure from employers and passengers, downsizing, encouraging senior level workers to quit, passenger violence or aggression, de-professionalization of the check-in job, reduction in wages and benefits, and lack of communication or support from management.

But work organization and technology can be integrated to promote worker well-being in any enterprise, in any country. Across different cultures, technology and job design should give priority to:

- Reducing excessive demands on time pressure;
- Developing action programs to reduce and eliminate heavy lifting and uncomfortable postures;
- Developing and implementing training programs to increase skills and competencies that are applicable on the job;
- Enlarging the workers' span of control over their work situation;
- Activating worker participation in planning job activities and routines;
- Improving feedback, recognition, and appreciation procedures [6].

Howbeit, it should be recalled that worker's health and safety is inherently an adversarial problem between managers and workers. To ensure the enactment of any of the recommendations made herewith, workers nonetheless need to get organized, and to be better organized to take more control to assure their own safety, well being and satisfaction at work. Evidenced by many of the research results referred to in this book, employer-initiated voluntary initiatives, "dialogue," or the "safety culture" -based themes proliferating in workplaces and countries today have not brought about significant improvements in work conditions or in workers' health, despite the many nicely packaged slogans to the contrary.

The impact of the check-in workers' job on the individuals performing it can vary depending on the design of the check-in counter and baggage-handling system at any given airport. The resulting workload and risk for MSDs from these exposures, can be likened to the MSD risks found in industrial workplaces where heavy manual lifting is involved. Workers in industrial workplaces doing heavy manual lifting (most of whom are men) do not, however, have to wear pumps, i.e., pretty, feminine little work shoes with a nice heel and the foot exposed, nor do they have to wear skirts and dresses while performing manual lifting, or while climbing over baggage carousels, check-in scales and conveyors, or while straddling stools provided in place of chairs.

The lack of a consistent framework for classifying injury and illness data means there is inadequate statistical information available. To establish the extent of occupational injuries and diseases and their cost among check-in workers, it would be necessary to approach all airlines and ground handling companies individually, since they are the only ones having reliable statistical data. Here too is a call to action for workers and their trade unions.

The higher prevalence of upper and lower back pain at Airport A in Ontario, compared with Airports B and C in Québec and Switzerland, provides evidence supporting the need for fully mechanizing baggage check-in systems. Eliminating the requirement that workers lift and carry each and every checked bag is an important first step to prevent work-related MSDs among check-in workers. Factors that need attention include unequal work load throughout a shift; lack of training and skills development; high levels of stress due to high work demand and a low level of worker control; daily tension from potentially aggressive passengers with a lack of empowerment-based worker training and management systems to protect workers; lack of communication with management; lack of management recognition of check-in workers' skill contribution for passenger, crew, and aircraft safety; the need for adjustable work stations designed for workers' requirements; and the basic need for worker participation in decision-making.

The findings from this study suggest that the physical and psycho-social stressors inherent in check-in work have a direct negative impact on workers' personal lives as well as on their productivity and efficiency at work, besides causing absenteeism from work. These demonstrated results have hidden yet real costs to both workers and employers, though employers often succeed in externalizing the costs.

Check-in Workers:
Safety Professionals on the Ground

I have suggested that check-in workers' skills and occupational niche be expanded and enhanced so that the employees can play a recognized role as safety professionals in civil aviation. The check-in worker's job naturally lends itself to being combined with the role of safety professional on the ground. In expanding the job, it is important that check-in workers have the confidence, commitment, and authority to deal with aggressive and disruptive passengers. Check-in workers are the logical choice to assess a variety of passenger factors associated with ground and air rage (assessing the weight of bags, ensuring that carry-on baggage is of the correct weight, size and number, identifying passengers who are under the influence of alcohol or exhibiting mood problems at time of check-in). Skills development and role enhancement also can offset the repetitive and cyclical nature of some of the job tasks, an important additional benefit. Given the necessary training to empower them to take on this critical role with management support, check-in workers are well placed to identify potentially aggressive or threatening passengers and to catalyze a chain of actions aimed at preventing disasters or aggressive behavior towards check-in and other air transport workers.

Limitations of the Research

Participatory Action Research

Participatory action research (PAR) is not without limitations and difficulties. One of them is to ensure that all groups understand the process and feel validated and valued in their contributions to the research design, process, and any outcomes. It requires that researchers speak the language of the various participant groups and not use a strictly academic approach in communication. PAR researchers must be attentive so they do not speak only in academic terms with the various groups, so they do not presume the same understanding of research terminology and of interpreting data, but without appearing condescending or technocratic. Academic jargon in PAR carries with it the potential to alienate the various participants in the study, who may not understand the concepts but who may not wish to appear less knowledgeable than others in a given discussion. PAR researchers must maintain a delicate balance: they should see themselves as learners and facilitators at the same time. Listening to and learning from workers and managers about workplace issues is valuable and enriching, though time-consuming.

The researchers have the additional responsibility to help the participants learn from the process. This teacher/facilitator role is time-consuming; it often involves working with groups in evenings, on weekends, at times when their own groups (management or union) can be organized to meet. Considering these obligations of the PAR process, purely quantitative research, defined, implemented, and interpreted by the researchers, without consultation and agreement from the participant community, is both easier and faster.

Another difficulty in applying a PAR approach is arriving at consensus with workers and employers. Doing so can be both difficult and lengthy in process for the researcher. It is more time-consuming than research based on the professional expert model, for example, where input from various groups is not involved and where consensus-building is not required at any stage. The consultative process in using a PAR methodology is likely therefore to be more costly in terms of researchers' time and in analysis of qualitative data. The study takes longer to carry out than when the researcher maintains complete control over the process and implements as s/he wishes. An additional time-consuming aspect is establishing a relationship with each group and bringing them into the process.

The risk of researcher bias through personal involvement with the participant community, as well as in any change process that may develop, is an additional potential disadvantage of PAR. The researcher can influence how the change process unfolds, how the research findings are interpreted and applied within and beyond the participating groups. Care must be exercised to maintain a distance from the process, allowing the stakeholder groups to define how a change process is envisioned and formulated. In the end, it is the workers and managers who must live with the effects of any changes they implement or any they do not implement, since non-implementation of changes, after awareness has been raised, can have consequences.

A noted shortcoming of using a PAR methodology is that priorities of any participating group may change during the process. This possibility can influence the degree of support to the researcher during the study. The researcher may well wish to study a particular issue, but doing so can become an obstacle if the issue is not considered a priority for all participating groups. If support is lost for any reason, the research team even may be obliged to abandon the study.

Questionnaire

The study group conducted one round of questionnaire distribution because of confidentiality and anonymity to protect the workers, which meant that non-respondents could not be tracked if a second round were to be conducted. A consequence of this decision was that a lower response rate had to be accepted than might have been achievable with repeated surveying. The study group accepted the response rate in order to ensure anonymity and to maintain an un-pressured environment for workers' and managers' input to the study. In particular, it did not seem advisable to pressure the management to engage in a second round of questionnaire distribution.

One potential limitation of the self-reporting of symptoms is that more workers with something to complain about may self-select to answer the questionnaire, compared with workers having no symptoms or complaints. At the Swiss airport, for example, there were two employers of check-in workers, one of whom paid lower wages than the other. Workers employed by the low-pay company performed check-in work exclusively. Check-in work under the other employer was combined with other duties, such as gate work or assisting passengers with special needs. In principle, workers were not meant to perform check-in work for more than four hours at a time. The employer paying lower wages and offering less desirable conditions of work appeared to prefer hiring young inexperienced workers. Turnover was fairly regular and workers indicated in interviews that they did not expect to remain on the job more than a year or two because of the low wages and boring, repetitive work. More workers of this employer might have self-selected to respond to the questionnaire than workers employed by the better paying firm since they may have had more' complaints about their work conditions. The employer paying better wages employed more check-in workers than the employer paying lower wages. Responses from the airport in Switzerland were aggregated. Data aggregation plus questionnaire anonymity meant that it could not be determined which management company was the employer of a given respondent.

A variety of non-work-related activities can cause or aggravate existing MSDs. Activities that workers engaged in outside of their job were not explored through the questionnaire because where both work-related and non-work-related contributing factors might exist, direct causality of MSDs is difficult to establish. Indeed, for a first-phase, exploratory study of an occupation not previously investigated, the study group considered it sufficient initially to examine and attempt to quantify only the work-related exposures. This study was not an attempt to establish irrevocable causality between symptoms and exposures. Within the framework of an exploratory study, the objective was to try to demonstrate and quantify exposures, symptoms experienced, and indicate some possible associations, while suggesting areas for further research.

Self-reporting of symptoms was subjectively based on the feelings and perceptions of the participants. While non-work-related activities might have contributed to MSD symptoms, workers might not necessarily make an etiologic association between MSD symptoms and activities they engaged in outside of work, which could cause or contribute to their pain. Workers might not make a direct association between symptoms experienced and work-related exposures, even when on-the-job exposure time meant that a particular position was adopted for many hours each day. It is possible that workers might experience musculoskeletal pain that is induced at the outset by non-work-related activities, but which might worsen, even to the level of a disabling injury, due to a number of work-related factors. The reverse could be true as well, with job factors triggering a problem that might then become aggravated by non-work-related activities. Thus, while exploration of non-work-related exposures was not within the framework of the present study, it is important to remain open to causal associations between work-related exposures and symptoms among check-in workers. Notwithstanding the length of time spent on the job, the repetitive nature of the work performed and the regularity of work (performed daily throughout each year) normally added up to more exposure time, and in many cases more body load, than what is caused by most off-the-job activities.

In the study results, upper back pain may appear to be less than it is in reality (29.8% of respondents reported upper back pain) because respondents may have been reporting pain in the shoulder area when it should have been counted as a self-report of pain in the upper back. Even though shoulder pain was a separate question, it might have been difficult to distinguish the upper back from the lower shoulder area, for example, when reporting on pain in the general area of one or the other, particularly if one had pain in both areas. What would appear to be important from the findings is that significant numbers of workers suffered from shoulder pain, upper back pain (with or without lower shoulder area included), and lower back pain.

While it would have been useful to make a clear gender-based analysis of the study results, it was not possible due to the very small number of male workers at each study site—too small to draw any gender-based conclusions.

Questions directed at management practices related to work demand, speed of work, organization of shifts, inclusion of non-check-in duties in check-in work, and an airline's quick-turnaround policies might have been useful to include in the questionnaire. However, focus group discussions indicated that questions of this nature might have been viewed as provocative or antagonistic to management, risking loss of their support for the study. Thus, obtaining information on these issues through interviews and in direct discussion with union representatives and the human resources departments, where possible, was preferable.

The formulation of questions addressing lifting and carrying baggage did not seem to differentiate sufficiently between manual lifting and carrying baggage and handling baggage during typical operations such as sorting out baggage jams. Postural discomfort and complaints of MSDs among check-in workers at Airports B and C in Québec and Switzerland appeared to result, at least in large part, from prolonged sitting or standing and awkward body movements, rather than from manual baggage lifting. This interpretation is supported by the work observation and videotaped work analysis conducted that revealed the awkward postures adopted by workers at Airports B and C. Observation of the work process further showed that indeed the fully mechanized baggage-handling system obviated the need for workers to lift and carry bags frequently, although it did occur, particularly when bags fell off the conveyor.

Interviews

In conducting interviews, the potential existed for bias to be introduced by those asking the questions (the listeners), as well as by those answering the questions (the respondents). Listeners could influence a response by asking a leading question or by clarifying for respondents according to their own perception. Either possibility could alter the way a respondent "heard" a question. In an attempt to avoid any such bias, structured questions for each study site were used for standardization purposes, but without strict wording provided to the listener. Because workers and supervisors were interviewed in two different countries, it was determined useful to allow the researcher flexibility in the phrasing of questions, as long as the question was focused on the item listed. While English was spoken by all workers and supervisors interviewed, at

Airport C in Switzerland, for example, English was not the first language of most of the workers and the supervisors who were interviewed. Interview topics were therefore structured in a list, allowing some flexibility in phrasing to optionally simplify some words or provide synonyms as needed.

To decrease bias during interviews, the researchers were careful to not interject personal thoughts or observations into the interviews; only the response to the questions as presented by the respondents was recorded. Notes from all interviews showed that the question items were clear and straightforward and offered no confusion for respondents. Subsequent to the airport visits, the researchers met with each other to discuss and re-review observations, new ideas, and questions that had come up, and to re-describe among the researchers the characteristics of the workers that had been observed. Discussing in groups aimed to decrease bias from inflated negative aspects that might have come up, particularly during interviews or through open-ended survey questions.

Pilot Testing

Ideally, the questionnaire and interview schedule would have been pilot tested in an airport that was not used as a study site. While less than ideal, it was necessary to pilot test both instruments in one airport where the questionnaire and interviews later were administered to all staff. In conducting research in real-life workplaces, one is often at the mercy of mergers and acquisitions, which sometimes can throw a wrench in the works. The process of negotiating for employer approval to conduct the study at each airport was lengthy and time-consuming. The process had been underway with the airline employing check-in workers at a major international airport in Canada. The airport was desirable due to its size and the fact that only the international terminal used a fully mechanized baggage check-in system, while the rest of the airport used a semi-mechanized system. When the negotiations were nearly finalized, management announced that it was about to acquire the other major airline operating out of that airport. This plan led the airline to withdraw from engagement in the study, due to delicate negotiations and company restructuring resulting from their acquisition.

It had taken six months to negotiate between the union and management to try to obtain the approval for that particular airport as a study site. When the airline management withdrew from engagement in the study, the search began again to identify a comparable large international airport, where the research group could obtain entry and support for the study in agreement with union and management. An additional difficulty was the study design, which demanded three airports of different sizes and check-in work station conditions. Access to Airport A, a small, regional airport, had already been secured when Airport B fell through. At the same time, the research team was searching for a third study site airport. Securing a replacement site for the original Airport B took an additional six months.

No workers at Airport A in Ontario mentioned that they had already completed the questionnaire or answered the questions; indeed, they did not seem to remember the pre-test. The pre-test was done discreetly, while full implementation of the interview and questionnaire with all workers was done with full introduction by management and repeated researcher visits.

Compensation Records

Examination of workers' compensation records proved difficult and problematic in both Canada and Switzerland. This part of the methodology had numerous limitations, such as the unexpected acquisition of one major airline by another at the time the original Airport B was in negotiation as a study site. This development precluded us from gaining access to compensation statistics from the airline employing check-in workers at that airport.

The study team wanted to examine Canadian and Swiss workers' compensation records in order to compare the national data with the airline or airport management company records. The researchers wanted to know whether it would be possible to obtain a picture of lost work time due to injuries among groups of workers comparable to check-in workers (computer workers, supermarket workers, airport baggage handlers, or general service-sector workers), and types of work-related injuries that were compensated, particularly work-related MSDs. One could then compare the national figures with the airline management's records to know how check-in workers' reported injury rates and lost work time rates compared with similar groups. This part of the study encountered unforeseen circumstances.

Occupational health and safety statistics in Canada are complex due to the lack of a consistent national framework for classifying injury and illness data. Since each of the ten Canadian provinces and three territories is responsible for collecting its own data, it is not possible to collect and disseminate complete, integrated national statistical information. For now, workers' compensation records in each of the provinces and territories are the most valuable sources of statistical information on work-related injuries and diseases resulting in lost work time. These statistics, however, are difficult to access and evaluate because of variation in coding systems and lack of standardization between provinces and territories. Each of the thirteen Canadian jurisdictions has their own occupational health and safety act and regulations. Each jurisdiction also has its own system of dealing with the adjudication of injury and disease. Between Canadian provinces, workers' compensation programs have differences in benefits payable, administrative practices, and the compilation of statistics. The data collected from the basic compensation forms are geared toward the adjudication of claims for traumatic work injuries with no uniformity or consistency because different workers' compensation boards use different coding methods based on their own needs.

The Association of Workers' Compensation Boards of Canada (AWCBC), an umbrella organization representing thirteen member jurisdictions, collects data from provincial and territorial workers' compensation boards and commissions. The AWCBC summarizes the information on all injuries and diseases that caused lost work time and that were accepted by the boards and commissions and publishes reports under the title *Work Injuries and Diseases*. In the present study, the most recent report, for 1996-1998, was examined. Although the report lists lost work time injuries by occupation and industry, only general categories are provided. In seeking more detailed information from the AWCBC, it became clear that only statistical data about airport personnel in general would be obtainable. The AWCBC did not have a specific code for reporting loss time injuries among airport check-in workers.

Statistical data about occupational injuries and diseases and data on lost work time among check-in workers are, in general, maintained by the employer. The employer might be an airline, an independent management company, a finance company, or a local airport authority, varying from one airport to another.

Because of the impending bankruptcy by the major airline in Switzerland, the major employer at Airport C could not obtain the desired workers' compensation data for Switzerland. To do so, they would have had to contact their major employer—which was the airline about to declare bankruptcy—and the Swiss National Insurance Fund. Politics. Mergers. Acquisitions. Bankruptcy. Participatory research can be complicated.

The lack of a category for MSDs in Swiss and Canadian workers' compensation data meant that even if one were able to obtain the compensation data from the central insurance fund or board, one could not identify work-related MSDs. The need to investigate the real costs to employers associated with the design of work processes, work organization, or work station design for check-in workers is punctuated by the significantly different lost work time injury reports provided by the two different employers of check-in workers at Airport C in Switzerland. It was explained in part by the fact that one employer encouraged the hiring of young, inexperienced workers, paid lower wages than the other employer, and accepted both high turnover of workers and shorter employment duration.

Still Working in the Dark

What is clear is that not enough is known about associations between jobs and workers' health through job analysis. I consider it a limitation that the mainstream approach in North American ergonomics and occupational health is to examine work-related illness and injury as a direct result of work conditions. Jobs exist within, and as part of, entire systems comprised of a variety of factors, each of a very different nature. Physical factors, organizational factors, psychosocial factors, and characteristics particular to a worker group all affect a job.

Work-related health problems generally result from the degree of interaction between some or all of these factors. The interaction can result in negative health outcomes in workers while some elements of a work "system" may act as buffers against negative effects. Trying to untangle the impacts of these independent factors is no easy undertaking. Work is complex because humans are complex.

Some researchers describe negative impacts on workers' health as coming from the lack of worker participation in organizational decision-making. Others describe the impact of control over workers and its relation to health. Many studies are worker-centered, rather than work-system centered. Worker-centered approaches often result in workers being the focus of behavior change instead of focusing change strategies on entire work systems, including work design, work organization, managers, and management approaches to workers' participation in decision-making. Most researchers look at physical factors as the dominant forces contributing to MSDs, though some (few) add psycho-social factors, particularly psychological distress, as variables [8]. Even fewer look at the impacts of *dehumanizing managers* on workers' physical and mental health, particularly the chronic psychological distress that such managers engender and the long-term effects and costs to subordinates and employers.

Indeed, it is important to avoid technocratic approaches in issues related to workers' health and well-being. This means avoiding slipping back into the paradigm of the past, which still dominates the field of occupational health, witnessed by the too few researchers who consider their role as including an approach oriented at change, and keeping the occupational health agenda focused on social justice—in all countries.

> Documenting the miserable working conditions in most developing countries is necessary but not sufficient to put occupational health in the limelight. Occupational health researchers should step out of their technical/disciplinary isolation and integrate their research within a wide social justice agenda. Hence, examining the impact of working in hazardous conditions on the health of working children, women, and men is not an aim in itself but rather a means to improve these workers' livelihood and quality of life. This cannot be achieved unless we join hands with researchers from other disciplines, labor unions, and community organizations who share the same objective [9].

A purely technocratic view leaves large holes in what we need to know, to enable us to move toward more human ways of working. Research conducted from a technocratic perspective makes it difficult to obtain evidence that workers can use to negotiate protective collective agreements or on which better and more protective legislation can be based. And technocratic approaches, particularly research looking at only one or two factors in a work system, have not contributed greatly to creating a knowledgeable and empowered critical mass. The interactive nature of many different elements in any work environment and the need to investigate from this perspective is so self-evident that it has even been said that studies examining only one variable related to MSDs should not be considered conclusive [10].

The need for critical mass cannot be overemphasized. "Occupational health researchers in developed countries sometimes achieve policy results without advocating and without entering into alliances with other groups. This does not mean that they might not achieve better results if they did enter into alliances . . . [with] social and political scientists, labor unions, and community groups" [11]. Participatory Action Research methodology is a useful tool for catalyzing such alliances, within the natural framework of a rights-based approach to workers' health, an important focus of a social justice agenda.

Trade unions have a critical role to play in working with researchers as well as in worker education, organizing for change, and managing change for workplace health. Howbeit, the role of unions in protecting workers' health and well-being needs to be amplified and strengthened, for example, through the mandatory inclusion of trade union health and safety specialists, independent of those pro-vided by management. If worker (and management) education and awareness-raising are to be taken seriously and are to have an important change-oriented impact leading to improvements in work conditions, then union-based or joint union-management health and safety committees should be instituted in every place of work. In past decades, these have been on the decline in many countries, both in number and in strength where they do exist. Union-led worker education courses have a long history of success in many countries around the world, as do union-based study groups, such as those developed in Scandinavia. Trade union-led education courses and study groups are a logical forum for establishing research priorities and generating critical research questions, which can then be discussed with researchers who are ready to work with unions, and management, through participatory action research.

It is that critical mass, empowered with knowledge and understanding, that needs to use its "voice" to raise the bar, if unacceptable work conditions are to change to the acceptable. There may be more information available today than ever before, but still the "powers that be"—those in decision-making positions—demand irrefutable evidence about the impacts on workers' health from different factors, in particular, the impact of management practices, psycho-social factors, and characteristics associated with women's work. Another gaping hole in avail-able information concerns the impacts on workers' health from positive influences, such as relations with management, social support, and worker control, which may interact to decrease the existence of or the effects of MSDs. An approach that looks at multiple factors is essential when examining multi-factorial problems. And the need for workers to take actions themselves to improve work conditions and not accept the unacceptable is crucial.

Overall, there is a glaring lack of research on the gender influences in MSDs, common in computer work, supermarket check-out work, and airport check-in work, all predominantly women's jobs. Where studies do exist, such as those on airport baggage handlers, the findings are not extended to other jobs having similar characteristics, such as relevant women's jobs that have not been studied.

Evidence shows that psycho-social factors, such as violence, emotional labor, and interaction with the public, are causes of high levels of emotional distress in workers. But very little information exists about the impact on workers' health, and the relation to MSDs in particular, from the psycho-social factors. Even less is known about how psycho-social factors contribute to MSDs where there is no management support system at work. One of the main problems with all of these information holes is that they create a widespread impression that a range of jobs are "safe" or "clean," without any particular danger to the workers performing them, in particular jobs performed mainly (or exclusively) by women. When we do not know about particular jobs, nothing is done to improve them or to protect those who perform the jobs.

Exacerbating the lack of information in today's unprecedented "information era," far too little research has been conducted on the relation between jobs and health outcomes using methods that combine knowledge generation with intervention and improvement. More worker-based research would be useful and is needed.

This exploratory study provided new information, but we still operate relatively in the dark in many areas. (The more you know, the more you realize you still do not know.) Shedding light on some of those dark areas would promote more understanding about systems that may have negative impacts on airport check-in workers around the world and what needs to be done to change the unacceptable.

Where More Light Would Be Useful

The exploratory study was conducted leaving open the possibility of later conducting a larger more in-depth study based on the findings. Results of this study can be applied to check-in systems of any design and level of mechanization. Recommendations can be made based on the new knowledge obtained and can be discussed with workers and managers. Whether the results of the present study could be used to establish a benchmark for making improvements in the check-in job could be investigated. Where management wanted to identify the problems of a specific worker group, the questionnaire and/or the interview schedule could be re-applied.

Similarly, the baseline of information established by the present study on the range of problems experienced by check-in workers, as compared with computer operators, supermarket check-out workers, and airport baggage handlers, provides a body of knowledge from which many concrete solutions can be implemented. Such solutions should be aimed at protecting workers, reducing lost work time, and improving productivity levels for management.

For example, basic principles about work organization, job enrichment, job enlargement, worker control, and ergonomics can be applied across different cultures in different countries. Indeed, they have been studied in various cultural contexts. This said, no information is available to tell us how airport check-in technology is transferred to developing countries. Check-in systems that are designed in one region and introduced elsewhere may introduce problems where workers are smaller, for example, or where adjustability is not built-into work stations. In transferring technology from industrialized to developing countries, very little (if any) consideration is given to potential negative effects on workers from the changes. To be marketable, most technology designers, producers, buyers, and users consider only economic and technical factors in their decision to produce or use a technology.

Before weighing economic and technical aspects, technology must also be considered from other points of view, for example, ensuring that it will be ergonomically, socially, and culturally sound [12]. In today's globalized world where production and services for companies in one country may take place in any number of other countries, gathering and exchanging information on workers in those other countries becomes more important. For example, if anthropometric details about workers were known in advance, technology designers could use that information to ensure that tools, processes, work systems, and so forth, would "fit" workers at the other end. Of no less importance, however, the education and training of designers, producers, and buyers in particular would have to impart the importance of obtaining and using such information.

A flexible standards-based approach could be useful in airports in developing countries where national health and safety legislation may not exist and where awareness of ergonomics may be limited or non-existent. Negative health outcomes can be prevented if the need for adjustability and local work conditions are taken into consideration at the design stage. Equipment and work systems that seem to work in industrialized countries may not always be appropriate in developing countries. Workers may have different physical and cognitive features in different countries and the way in which the features interact with a particular job can vary. For use in developing countries, it is important that the configuration of airport check-in work overall is made in the local context in which the work is done, for acceptance by, and good fit with local work forces.

In developing countries, some interesting and inspiring examples have been shown of successes in reducing physical load and improving work organization and scheduling, using relatively simple changes, that bring important and positive results to productivity and worker well-being [13, 14]. In one example, comparable industrial workers in Indonesia and Sweden considered the same aspects of job design to be important, and the findings showed that the implications for technology and job design were the same for both countries [15]. The use of simple solutions is relevant for the prevention of MSDs in countries of vastly different stages of industrial development. Such findings can be extended to any airport check-in situation.

The reality of how enterprises apply the findings of research, such as those in this study, is an open question. Research may or may not facilitate change. Through this study, I hope that the participatory action research process has helped to produce new, directly applicable knowledge for both workers and managers. "We need not more highly trained and sophisticated researchers operating with ever more esoteric techniques, but whole neighbourhoods, communities, and nations of 'researchers'" [16].

Outsourcing a process is one approach used by many companies to relieve themselves of the responsibilities and initial costs stemming from change in work organization, work process, or disability and absenteeism associated with the way work is designed. While it merely transfers the problems to another company and to new forms of management, in reality the problems no longer exist in relation to the initial employer from which the job was outsourced. In today's world of fierce competition, a company winning an outsourcing contract by offering the lowest price may not view investment in healthy work design as cost-effective. A long-term vision on investment-return is needed, and this too becomes an open question of approach for companies competing for contracts by cutting costs to the core.

This raises an inherent question about the long-term social viability of the social and cultural hegemony, the dominant economic model, with its labor market incentives for outsourcing. Where companies do not have an inherent ethical approach of incorporating workers' health into their management practices, regulation takes on greater importance as a means of holding contractors and subcontractors responsible for work conditions and worker health. More debate is needed to demonstrate to employers the costs of "negligent" practices. The existing body of evidence lacks breadth of occupations, most notably in women's work.

It would be very useful for workers, employers, trade unions, health professionals, and practitioners to know more precisely the costs to check-in workers' employers and insurance companies that stem from MSDs and lost work time due to MSDs, worker turnover, training and re-training. Trade unions, in particular, have a critical role to play in identifying the extent of anxiety and psychological distress from incidents involving aggressive passengers or from abusive management, as well as the costs related to the psycho-social factors associated with check-in work. Who else is likely to take the lead?

It would be equally important to know more precisely the indirect costs associated with those MSDs that are not registered officially with employers and to have a clearer picture of indirect costs to workers, their families, and employers from reduced efficiency, reduced productivity, and lowered morale.

This kind of cost-related information could provide a convincing case for management to justify making the needed changes. Equally important, this kind of cost-related information is powerful for collective bargaining.

As mentioned above, there is a glaring lack of information on gender-related influences in MSDs, common in women's computer work, supermarket check-out work, and airport check-in jobs. Where studies do exist, such as those on airport baggage handlers, the findings are not extended to other jobs that have similar characteristics, such as relevant women's jobs that have not been studied. Therefore, knowing more about the gendered dimensions of check-in work, particularly information about strength testing and maximum allowable weight limits in different countries, with a particular focus on the effects of lifting weights on women's health, would be helpful. In its conclusions, the 1996 International Congress on Women, Work and Health highlighted the need for more appropriate gender-sensitive methods for research and prevention with the direct participation of women workers. To reach such a goal, closer networking of researchers, experts on women's occupational health issues, and trade unionists is necessary [17].

There are other areas about which we are still relatively in the dark, which, if illuminated, would help us understand more about why workers in various occupations accept the unacceptable. Such information could be extremely empowering for all concerned, but particularly for contributing to critical masses of voice, working to raise the bar.

Many airport workers all over the world have reported the same thing: once you have worked in an airport for five years, you cannot work anywhere else—it becomes part of you. Many check-in workers said that they intended to leave their job, either soon or in the foreseeable future. It is the final response of those workers to various elements of check-in work that they find unacceptable. Most of the workers who plan to leave their jobs had not yet been on the job for five years. Perhaps working in an airport after at least five years is so attractive and interesting, creating an aura of status and glamour associated with flying, that it might even cause check-in workers to accept their work conditions and related injuries. Knowing more about the positive psycho-social elements associated with airport work might help to change the trend that sees workers registering injuries far less frequently than they actually experience them.

The positive impacts of social support are sufficiently convincing to raise additional questions related to airport check-in workers. Would enhancing social support and positive psycho-social factors in conjunction with ameliorations in work station design as agreed with workers, together with management measures aimed at re-professionalizing check-in work "balance" the negative job factors enough to lead to a reduction in MSDs? "Before" and "after" measures of job satisfaction would complement such an investigation.

Because psychological stressors and physical strain may vary over time, a job stressor related to worker strain at one time may not be related to worker strain at another time [18]. Therefore, a prospective study collecting repeated data on worker strain, as well as on job design, might be useful to show whether lost work time increases over time among the same check-in workers.

The issue of violence toward check-in workers is critical and growing. While action has concentrated on air rage to protect crew and passengers in the air, more information is needed to identify the overriding cause, or causes, of ground rage. Discussions with airport management at the three study sites and with airport workers from around the world always arrived at the issue of ground rage; everyone gave their thoughts on the main cause(s). Many check-in workers said drugs and alcohol were the main factors in ground rage. Others felt that increased psychological stress, fear and anxiety associated with travel were the cause, while still others believed that more passengers arrived late at the airport and felt out of control when traveling. Are airline advertising promises, which are meant to create customer loyalty, a direct cause of passenger violence or aggression toward check-in workers? Do such advertising promises create a feeling of entitlement among customers, to the point that passengers feel justified in behaving abusively to check-in workers? To what degree do perceptions of entitlement, created by advertising, occur in other service-sector jobs? Is manipulation of the masses through advertising causing increased violence toward service-sector workers in general? Most likely, a combination of factors creates the guilty culprit, but the weight of the effects of airline advertising promises should not be underestimated. Because the question is as yet unanswered, at two international trade union health and safety conferences in four years, ground staff and cabin crew union representatives asked urgently that the question be studied so that they could better know how to (re)address policy recommendations.

Management policies are needed to prevent violence against workers and to deal with the aftermath of violence when it occurs. Here again is a call to action by trade unions. Evidence of a problem is needed to justify pro-active and protective management systems in jobs such as airport check-in work, as for many other service-sector jobs. The lack of knowledge about check-in work (and many other service-sector jobs) is an important cause of the lack of evidence needed to justify pro-active and protective management policies. With no prior description of the magnitude of the problem in check-in work, management policies have not been developed to address the problem.

Little attention is paid to most service-sector jobs because they appear relatively "safe" on the surface despite numerous potential inherent hazards and because they are characteristically women's work. Obtaining more information about occupations dominated by women workers can help to dispel the myth that women's work is "safe" work. There is a great deal of similarity in many characteristics typifying women's work in the service sector today.

But what about service-sector jobs performed mainly by men, which have not been investigated? A similar approach could be used to gather information on many service-sector jobs that entail lifting, carrying heavy objects, and awkward postures, similar to the hazards faced by check-in workers. Examples of such jobs include airport shuttle service drivers and package delivery workers. Applying a similar approach for examining a variety of service-sector occupations would quickly help to build a body of information and evidence about work-related exposures and damaging management practices. Doing so could help develop recommendations for improvements in work conditions, leading to reduced costs for employers and increased efficiency and well-being for workers.

Research to identify the causes of work-related adverse health in workers is important. Identifying suggestions for improvements or solutions to known or suspected problems is equally important. But it is the joining of both strategies—identifying the root causes of problems and using that information to make studied suggestions for eliminating those problems—that is important for protecting workers' health and for improving productivity and efficiency.

To reduce the knowledge gap, we need more studies conducted with the participation of trade unions if we are to further understand the conditions in which women work and how policies, practices, standards, and equipment—designed for male workers—affect women's health, both on and off the job, and the relation between these factors and the existence of MSDs.

IMPACTS OF THE STRUCTURAL AND CULTURAL HEGEMONY

The field of occupational health and safety has not reached the stage of global advancement today that it should have attained, given the amount of technical expertise that exists in the field. Indeed, there have been many stepwise and important incremental improvements that have been made. However, before citing them, let us look no further than asbestos. There is perhaps more empirical evidence condemning asbestos as a substance that kills workers than for any other known substance. Yet political fights continue, with powerful industry lobbies wielding their force in international trade arenas arguing that asbestos can be used safely, even though all sound scientific evidence demonstrates the contrary. And a global ban on asbestos, which is the only sensible approach, seems a mere fibrous dream in 2006. If we are still at this stage over asbestos, with all of the supporting evidence in favor of a global ban, what can we expect for other work-related hazards, in particular those that are often argued as being more nebulous, like MSDs, work-induced stress, job strain, psycho-social hazards, psychological abuse, physical violence, not to mention the age-old well-known chemical, physical, and biological hazards? Clearly, stepwise, incremental changes are going to be with us for a while, as insufficient as they may seem.

If we consider the fact that a strong international community exists (albeit the occupational health community is not present in all countries or in all sectors, and is over-represented in a handful of rich, industrialized countries) with expertise in research and practice related to workers' health, work conditions, epidemiology, occupational medicine, industrial relations, and so forth, one could expect that by the 21st century, a rights-based approach to occupational health would dominate, superceding the traditional medical and engineering "silo" approaches, the paradigm of the past. Medical and engineering approaches, while immensely important and valuable, have been, and still are, dominated by "experts" and "technocrats." Some (if only a few) researchers work directly with trade unions, management groups, women's groups, NGOs, community health groups, environmental groups, and so on. Yet in reality, all major advancements in the area of workers' health have come from workers' own struggles. Will never the twain meet in a more holistic approach of respect and collaboration [19]? There is a huge unmet need for experts and researchers to work together with these groups, lending their technical assistance for social policy-making and for building agency.

Silo approaches notwithstanding, the lack of adequate advancement in the protection of workers' health is largely due to the major changes in the global economy since the 1980s. Politics, globalization, widespread labor market flexibility, new management practices based on work intensification, the push for ever-increasing productivity and the accumulation of capital, weakened collective voice, significant declines in the labor movement, moves from statutory regulation toward self-regulation, cutbacks in regulatory apparatuses and in state funding for the protection of workers' health, reductions in workplace safety committees and safety departments, and economic liberalization all can be fingered. In short, these causes reflect today's structural and cultural hegemony.

The regulatory changes and the market or self-regulation trend is based on the idea that complex comprehensive safety codes, large institutional mechanisms, and compensation for workers hurt, killed, or made ill by their jobs impose heavy costs on employers and create tendencies for workers to be careless (i.e., implying that workers would be careless because they can collect compensation for whatever might happen to them on the job). Requiring employers to pay heavy costs for injuries and illnesses of regular workers provides incentive for employers to employ non-regular workers and to use precarious contractual arrangements.

Taken together, the impacts of management practices on worker health point to a globalized and increasing lack of basic social and economic security, which induces workers to take a job no matter what the conditions of work. In today's economy, employers use employment insecurity, income insecurity, and representation insecurity (evidenced by diminishing union representation rates) as leverage when workers press for improvements in work conditions. If, for example, each and every citizen had the security of a minimum income as a guaranteed right, people would be able to make different decisions than those made out of fear and insecurity. A minimum guaranteed income, for example, would likely lead to improvements in working conditions: if employees did not feel obliged (out of fear and insecurity) to accept a job with poor conditions, employers would have to work harder to attract employees, shifting the balance of power. In many countries, there is a slow-building movement for a basic income as a universal citizenship right. For now, however, political support still seems sparse.

Health-breaking job strain induced by increased control over workers, the absence of worker autonomy in most workplaces, the lack of decision-making latitude by workers, long work hours, long and often arduous commutes, increases in workplace violence, work-induced depression, burnout, stroke, hypertension, and cardiovascular disease are factors and outcomes that cannot be ignored. Or should not be ignored. An employment contract is for performing a job, selling one's labor and/or knowledge, and engaging in decision-making. An employment contract does not include selling one's health with one's labor, nor does it include accepting abusive treatment or intolerable conditions.

In a system where the law of the strongest and the slyest reigns, abusers become kings. When success becomes the principal value, honesty looks weak and abuse seems resourceful [20]. This is the brave new world in which we live [21]. The answer to the societal problem that has taken root and which is growing rapidly has to be collective and cultural.

The challenge to trade unions to exercise leadership to change the cultural hegemony is great, more so given the global decline in trade union membership and strength. Nonetheless, there is an opportunity and challenge to stimulate widespread debate on re-conceptualizing work in the 21st century. Effective unions can lead in changing the cultural hegemony so that workers are put first—not managers and not capital. These challenges must be taken up both at the levels of the individual workplace and the society.

It is necessary to re-think collectively the concept of subordination, to clarify what is acceptable and what is unacceptable behavior in hierarchic relations between managers and subordinates. Workplace health and safety committees should include discussion about the manager-worker dichotomy, which should become less distinct. In healthy, solid work structures, in healthy organizations, *dehumanizing managers* cannot go far. If the enterprise or organization itself functions on genuine respect for its work force, it then has a positive effect on the individuals who make up that work force. Companies that respect their own work force attract and keep the best employees. This alone is the basis for the accrual of profit and for productivity [22].

Critical mass is imperative. Raising the bar through awareness and basic training would help to create a critical mass that is not willing to accept the unacceptable. Stepwise change can lead to a positive snowball effect [23], and collective action should be seen by top management of any workplace as an opportunity to introduce real change in management practices and to make improvement, not as something to fear. Success stories spread efficiently today, sent around the world electronically and in hard copy print. The Web linkages that keep people in touch around the world are powerful sources of information exchange, for building collective voice, and creating critical mass for positive change. The development of business womens' associations, women's caucuses, and women's groups in professional organizations and trade unions is a positive attempt to introduce change, particularly change in the power relations exercised in the workplace with the strong sex-roles and gender-based definitions that accompany and often define those relations [24].

Employers who give importance to treating workers with dignity and respect, to creating decent work conditions, and to encouraging workers to release their full potential and creativity have success stories to tell. Some of them write and speak publicly to share with others their "how to" stories and "good practices." It is important to circulate widely positive examples that do not necessarily correspond to the dominant socio-economic model. These often provide ideas, support, and courage to workers in far corners of the world who may struggle with destructive managers while looking for ways to protect themselves, and for alternative management practices.

In the face of the structural hegemony, new groups are making advances in various countries, particularly, but not only, in developing countries. Advances include innovations for the protection of workers in developing countries [25]. Merged with trade unions, these emerging "voice" groups could create a new and large body of collective voice with renewed strength. Alone, they stay as relatively isolated groups whose force still remains to be galvanized into coalitions. If significant change is to be made and major advancements to take place in the protection of workers' health and well-being, then coalitions are a necessary foundation for action and more sound social policy aimed at providing basic security to workers. Basic security must include sound work conditions and the protection and promotion of workers' health and well-being.

CONCLUDING REMARKS

This study of airport check-in workers has implications for organizational change, for social policy, and for generating improvements for workers. Efforts to improve any organization aim at increasing organizational effectiveness and improving the well-being of its employees.

In general, the amount of information available on what is happening to working people and on workers' health around the world is shockingly limited. Indeed, in most countries, the information that is available on trends in workers' health is often unreliable, often biased, or for various reasons not accessible. Employers, managers, and governments can do better. There is insufficient information available on airport check-in workers and inadequate statistical information about most jobs, due to the lack of a consistent framework for classifying injury and illness data, and because workers often do not report their work-related or work-suspected health problems.

Clearly new strategies are needed for dealing with well-known hazards and risks, such as those from dangerous substances, machinery, tools, and manual handling, as well as for emerging issues, such as psycho-social hazards and MSDs, and all of the adverse outcomes related to management practices. Most countries could do a much better job protecting and promoting the well-being of working people and applying management practices that do not engender ill health.

Management practices are not some abstract concept far removed from workers; they are developed, decided, and implemented by individuals. What makes work difficult is people. Without people, work essentially does not exist. People developed management practices that they believed would increase productivity and profit. The only laboratory where experiments are conducted on management practices is the real-world workplace. However, when various management practices are shown to cause ill health and other negative outcomes to workers, unfortunately those practices are not simply scrapped or labelled as some toxic chemical, "Bad Stuff, Dangerous To Human Health—Do Not Use."

In reality, workers perform a balancing act every day at work, trying to ward off and manage the negative psychological impact from poorly adapted systems of work, which include authoritative individuals in decision-making positions, de-humanizing hierarchy, and exclusionary decision-making. But what is the price to pay? I have heard workers say that they feel they are losing their sense of self by adapting themselves to *dehumanizing* management practices, including rigid hierarchy, or by trying to adapt themselves to destructive managers. In the extreme, is this not what has been linked to some people snapping and "going postal?" At the minimum, the spillover effects on people's lives outside of work are obvious.

Our heritage bears adverse effects from past methods of work. But we can decide whether to continue in the same vein or break with the past to make change based on what we know that those before us did not know when they adopted various methods of work. This is the essence of using individual and collective voice to bring about change, to exercise citizen and social responsibility, to not remain passive in the face of harm. It takes courage to stand up to the system of production and to be decent to and encourage workers. Courage can be hard to muster in the face of a giant. But collective inaction by workers means tacitly inviting and accepting being run over by a bulldozer.

People are smart enough to have invented all kinds of management practices and yet foolish enough to let certain of them continue to dominate our lives even when they should be shelved. The insidious increase in control by manipulation and the increase in numbers of supervisors and managers are strong indicators of an increase in negative forms of control over working people, with a decrease in workers' decision-making latitude and an increase in demands on workers [26]. Throughout the early 20th century, blue-collar workers were un-free to perform a possible task, bound by Tayloristic manufacturing processes, slaves to the automation generated by the industrial revolution. They had no degree of control, autonomy, decision-making latitude, or input to organizational decision-making. Today, even more workers are free to perform an impossible task, as modern day management practices create the illusion that working people are free to be creative, to perform, to participate, to engage, as management drives them to work faster and faster under conditions of chronic psychological stress and physical strain, under increasing insecurity, with no degree of control, autonomy, own-work decision latitude, or input to organizational decision-making. Surely we have the wisdom to banish destructive managers and management practices and all that goes with them. Is enough not yet enough?

Now in the 21st century, for the first time in history, we have a large body of knowledge that has analyzed the effects of various methods of management and production on workers and demonstrated undesirable and preventable costs to employers. The evidence points to the need to re-conceptualize how we wish to perform work in this century and, based on what we know, to re-think how we wish to define our relation to work. Are we obliged to continue to work in the way we do just because "that's the way it was always done?" As rational creatures, who think about the past, present, and future, we have the ability to change the direction of our futures, either socially or technologically.

It is time to set the bar sufficiently high as the benchmark for the treatment and conditions working people are prepared to accept in their place of employment. Any behavior, treatment, or conditions that fall below the bar should be considered unacceptable and should be treated accordingly. Scary stuff? Only at first. And we are worth it. If jobs were designed from the outset with people, not profit, as the priority, most jobs would not be designed and carried out as they are today. Even dirty jobs, such as mining, could have been designed differently, more humanely, had people and not profit, been put first. After all, we are paid to do a job, not to have the job "do us in."

> Whether 'tis nobler in the mind to suffer
> The slings and arrows of outrageous fortune,
> Or to take arms against a sea of troubles,
> And by opposing, end them.
> —William Shakespeare, *Hamlet,* Act III, Scene I [27]

ENDNOTES

1. Author's personal quote.
2. M. Kompier, The Psychosocial Work Environment and Health—What Do We Know and Where Should We Go? *Scandinavian Journal of Work, Environment, and Health, Editorial, 28*:1, pp. 1-4, 2002.
3. P. Froom, B. Cline, and J. Ribak, Disease Evaluated on Return-to-Work Examinations: Aviation Ground Personnel Compared to Other Workers, *Clinical Medicine, 67*:4, pp. 361-363, 1996.
4. T. F. Morse, C. Dillon, N. Warren, C. Levenstein, and A. Warren, The Economic and Social Consequences of Work-Related Musculoskeletal Disorders, The Connecticut Upper Extremity Surveillance Project (CUSP), *International Journal of Occupational and Environmental Health, 4,* pp. 209-216, 1998.

5. International Transport Workers' Federation, Air Rage: The Prevention and Management of Disruptive Passenger Behaviour, *Safety in Practice, 1,* p. 12, London, May 2000.
6. M. Hagberg and D. Wegman, Prevalence Rates and Odds Ratios of Shoulder-Neck Diseases in Different Occupational Groups, *British Journal of Industrial Medicine, 44,* pp. 602-610, 1987.
7. K. Ekberg, B. Bjorkqvist, P. Malm, B. Bjerre-Kiely, M. Karlsson, and O. Axelson, Case-Control Study of Risk Factors for Disease in the Neck and Shoulder Area, *Occupational and Environmental Medicine, 51,* pp. 262-266, 1994.
8. A wonderful and two-eye opening explanation of why mainstream research continues to be conducted as it is, despite the need for a shift in thinking and approach, is provided by K. Messing, *One-Eyed Science: Occupational Health and Women Workers*, Temple University Press, Philadelphia, 1998.
9. I. Nuwayhid, Lack of Occupational Health Research in Nepal: Nuwayhid Responds, *American Journal of Public Health, 95*:4, p. 551, 2005.
10. R. F. Bettendorf, M. S. Hoffman, E. Adams, H. McLoone, and C. J. Purvis, *The OERC Framework for Understanding Upper-Extremity Musculoskeletal Disorders Research, Theory and Management,* paper presented at The National Ergonomics Conference and Exhibition, New York, May 1998.
11. I. Nuwayhid, Linking Health to Social Justice: Nuwayhid Responds, *American Journal of Public Health, 95*:7, pp. 1090-1091, 2005.
12. A. Manuaba, Appropriate Technology Approach with its "Built-in" Ergonomics Aspect is a Must in Technology Transfer to Attain Sustainable Development in Industrially Developing Countries, in *Global Ergonomics,* P. A. Scott, R. S. Bridger, and J. Charteris (eds.), Elsevier, Oxford, pp. 57-62, 1998.
13. K. Kogi and R. N. Sen, Third World Ergonomics, *International Reviews of Ergonomics, 1,* pp. 77-118, 1987.
14. R. N. Sen, Application of Ergonomics to Industrially Developing Countries, *Ergonomics, 27*:10, pp. 1021-1032, 1984.
15. L. Waluyo, K. Ekberg, and J. Eklund, Assembly Work in Indonesia and in Sweden—Ergonomics, Health and Satisfaction, *Ergonomics, 39*:2, pp. 199-212, 1996.
16. B. L. Hall, Participatory Research: An Approach for Change, *Convergence, An International Journal of Adult Education, 8*:2, p. 30, 1975 (in M. Elden, Sharing the Research Work: Participative Research and Its Role Demands, in *Human Inquiry: A Sourcebook of New Paradigm Research*, P. Reason, and J. Rowan (eds.), John Wiley and Sons, Chichester, p. 266, 1981).
17. J. Paul, Women, Work, Health: A Gender Approach, in *Women and Health at Work*, Trade Union Technical Bureau Newsletter, 3, pp. 1336-1342, 1996.
18. P. Carayon, C-L. Yang, and S-Y. Lim, Examining the Relationship Between Job Design and Worker Strain Over Time in a Sample of Office Workers, *Ergonomics, 38*:6, pp. 1199-1211, 1995.
19. For a discussion of this issue, see E. Rosskam, Technical Assistance to the Grassroots, Part II: Erin Brockovich Revisited: A Lesson for Social Policy-Making and Agency, *New Solutions: A Journal of Occupational and Environmental Health Policy, 15*:2, pp. 107-112, 2005.
20. M.-F. Hirigoyen, *Stalking the Soul: Emotional Abuse and the Erosion of Identity*, Helen Marx Books, New York, p. 198, 2000.

21. A. Huxley, *Brave New World*, Doubleday, Doran & Co., New York, 1932.
22. M.-F. Hirigoyen, *Le Harcèlement Moral Dans la Vie Professionnelle: Démêler le Vrai du Faux*, La Découverte et Syros, Paris, pp. 397-398, 2001.
23. E. Rosskam, Technical Assistance to the Grassroots, Part II: Erin Brockovich Revisited: A Lesson for Social Policy-Making and Agency, *New Solutions: A Journal of Occupational and Environmental Health Policy, 15*:2, pp. 107-112, 2005.
24. N. Sokoloff, *Between Money and Love: The Dialectics of Women's Home and Market Work*, Praeger, New York, p. 243, 1980.
25. For detailed examples of innovative schemes from numerous countries, see chapter 13 in *Economic Security for a Better World*, International Labor Office, Geneva, 2004.
26. R. Karasek, *A Stress-Disequilibrium Theory of Cardiovascular Disease*, presented at the Fourth International Conference on Work Environment and Cardiovascular Diseases, Newport Beach, California, March 10, 2005.
27. William Shakespeare, Hamlet, Act III, Scene I, 1602, in *The Riverside Shakespeare*, Houghton Mifflin Company, Boston, p. 1160, 1974.

Epilogue

Since the completion of this study, a number of changes have taken place, all of which have been positive. Some have been direct outcomes of this participatory action research; others have been related to it, and some possibly have been indirectly linked.

Following dissemination of the study findings and discussion with the Canadian Auto Workers' Union, the local union together with management at Airport A in Ontario, Canada (the semi-mechanized airport where workers lifted and carried an average of 13,794 kilograms per day) re-designed entirely all of the check-in work stations in 2004. Photographs in the Appendix show the work stations before the changes and others of the work stations after the re-design. A fully mechanized check-in system was installed to eliminate the need for workers to lift and carry bags. This new system indeed will greatly help to relieve the physical stressors on workers' bodies. The union also is following up on other issues related to the entire work "system."

A public seminar to present the results of the study was held in 2003 in Switzerland. Participants included representatives of management from Airport C in Switzerland, check-in workers, union representatives, including a collective bargaining agent, labor inspectors, representatives from the national workers' compensation body and the regional Institute for Occupational Health, and the press. The seminar catalyzed action. The labor inspector responsible for work at Airport C met soon after with representatives of management and the union. A process of negotiation began using the results of this investigation to improve not only work conditions but also workers' employment and income security.

Check-in work stations at the Swiss airport were designed for sitting work only, which has been linked to MSDs in this worker population. Using the study findings and recommendations made in a report for the study participants, an increase in task variety has been introduced to relieve the Swiss airport check-in workers from performing check-in work exclusively. According to airport management in April 2005, check-in workers now perform different functions in the airport, including spending more time on gate duties, and workers are rotated more frequently. The introduction of greater and more frequent task variety reduces the time spent performing check-in work only and has a number of beneficial outcomes: body load from static posture is reduced; exposure to awkward postures and lifting and carrying bags is reduced; the risk of exposure to violent passengers is diminished; monotony is reduced; and workers have more flexibility and freedom during their work day. According to airport management, some days at Airport C are so busy that check-in workers do not even have time to go to the toilet. The introduction of this increase in task variety and flexibility in how work is organized and performed now enables workers to take a break when one is needed.

These changes will not necessarily eliminate all of the adverse outcomes discovered and described among check-in workers, but the changes are positive steps and hopefully will lead to further organizational changes in management practices. Change usually takes time, and more often than not is incremental and step-wise. Often, the very act of working toward and implementing one change leads to further changes over time. The empowerment aspect is worth noting here. By engaging and involving others in the research process, providing and discussing results and recommendations with them, those "others" have been able to use the research in their own way among themselves to make positive changes appropriate to their own work and organizational context.

In addition to the outcomes mentioned above, which are directly related to the research process, there have been a number of developments in other countries and contexts. A major airport in Scotland in 2004 introduced an airport-wide regulation limiting all checked baggage to 20 kilograms. Research on airport check-in workers in Malta was completed by a local researcher who applied the questionnaire developed for our study, including a body map. An ergonomist in Ireland is conducting ergonomics-focused research on airport check-in workers in that country and has made contact regarding our research.

In April 2005, IATA (the International Air Transport Association, which represents the majority of airlines around the world) passed an important resolution that limits all checked baggage to a maximum of 32 kilograms for all classes of travel. All airlines adopting the resolution will require passengers to send by cargo any and all baggage exceeding 32 kilograms. Passengers who travel economy class will still be able to pay excess baggage fees for total baggage weight above the 20 kilos limit, but only up to 32 kilograms. Business and first-class passengers will still be able to pay excess baggage for total baggage weight above their 30 kilos total limit, but also only up to the new 32 kilograms limit. Above 32 kilos, baggage for passengers in any class will not be accepted for check-in. The resolution passed because of the extremely high level of disabling back injuries among airport baggage handlers, a predominantly male population, and associated high workers' compensation costs. The adoption of the resolution by airlines around the world will clearly have positive spin-off effects for airport check-in workers, reducing greatly the amount of weight they will end up lifting and carrying each day. It is another positive step and a public recognition of the risks to one occupational group (baggage handlers), even if it was not intended for check-in workers and even if it was driven by the "bottom line."

The study prepared the groundwork needed for collective bargaining agents to take steps toward solving the problems identified among check-in workers. To get the information into the hands of collective bargaining agents around the world, in August 2005 the International Transport Workers' Federation published a report of the study (Rosskam, E., A. Drewczynski, R. Bertolini, *Service on the Ground: Occupational Health of Airport Check-In Workers*, International Transport Workers' Federation, London, 60 pages, August 2005) and disseminated it to the collective bargaining agents of the ITF's global trade union affiliates. The ITF represents 600,000 aviation workers in 110 countries. The report provides empirical evidence for collective bargaining where such evidence did not exist before. The report is available both by direct mail and on the ITF Web site. The Canadian Labour Congress and the Canadian Auto Workers' Union disseminated the ITF's report of the study to their affiliates and members at airports across Canada.

The ITF is considering extending their international media campaign against Air Rage to include ground staff as well as cabin crew. Using the study results, the extended campaign would aim to improve work conditions for ground staff in airports, re-professionalize the status of check-in workers, and prevent violence for check-in workers.

From the design stage of the study, both trade union and employer partners indicated their intent to make use of the results of this study. The study results have been presented and discussed in international trade union health and safety conferences. Aspects of the study have been presented upon request in seminars at a variety of universities, in scientific conferences, at social policy conferences, to various university MBA and management classes, and in public seminars for employers and the general public. The study was discussed at the ILO's *Tripartite Meeting on the Civil Aviation Industry* in January 2002. Conclusions from that meeting call for the ILO to continue research and follow-up work from this study of check-in workers.

Articles about the study have appeared in trade union magazines, ILO publications, journals, and in various newspapers, including a full-page article on the front page of the newspaper *Le Courrier*, August 26, 2003, in Geneva, Switzerland. The ILO Department of Communications published a feature article on the study in the ILO's magazine *World of Work*, No. 44, October 2002. The magazine has an estimated worldwide readership of between 30,000 and 50,000 and appears in English, French, Spanish, and German. Adapted versions of the feature story also were published in Russian, Chinese, Norwegian, Japanese, Danish, and Hindi. In January 2004, the Canadian Auto Workers' Union (representing airport check-in workers across Canada), published *Ergonomics in the*

Work Environment: A Manual for Workers, which includes a section on airport check-in workers with findings from the study and photographs of workers from the three study sites. The manual has been distributed widely by the CAW.

The ILO's Socio-Economic Security Programme published in 2003 a report of the study under its SES Paper Series. The report was made available to the ILO's major constituent groups: employers, workers, and governments, as well as to a broad range of academics world-wide, particularly social scientists in both developing and industrialized countries, and to a variety of Non-Governmental Organizations. The report is available on the Programme's Web site, in PDF format, and in hard copy by request. The CCOHS distributed the ILO's report of the study among its tripartite constituents.

A short video was made using film footage from the three study sites for use at workshops, meetings, and international conferences. The video shows check-in workers performing their jobs at the three study sites, highlighting the risks of MSDs. The video, available on a CD-ROM, has been provided to the ITF to distribute among its affiliates.

Appendix

**Photographs of Check-in Workers at
Study Site Airports, in Two Countries**

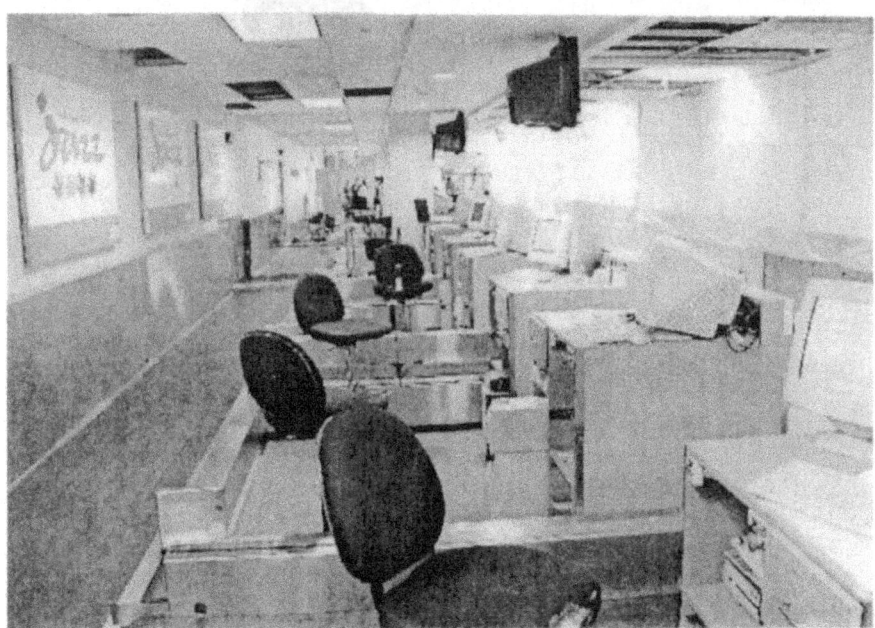

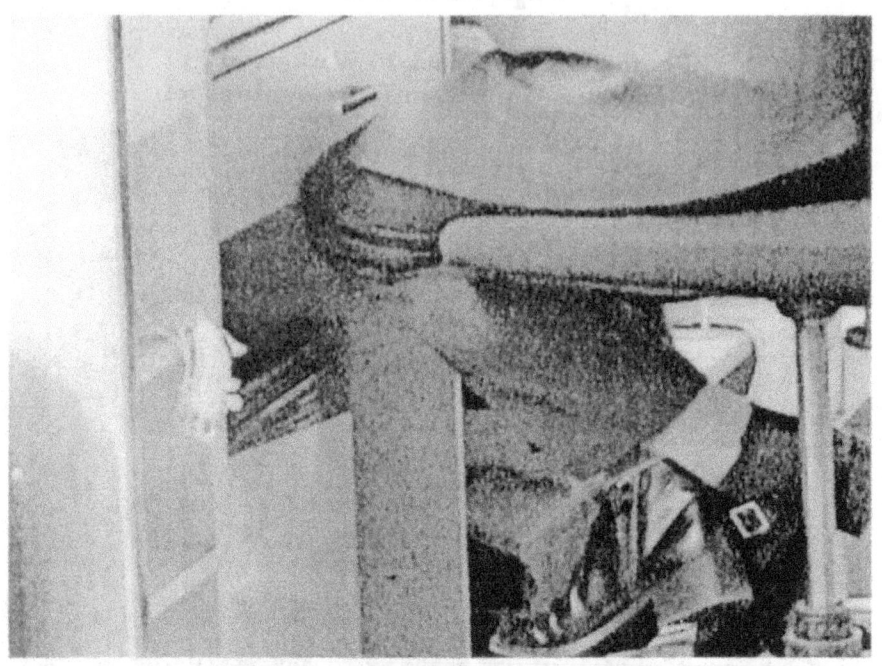

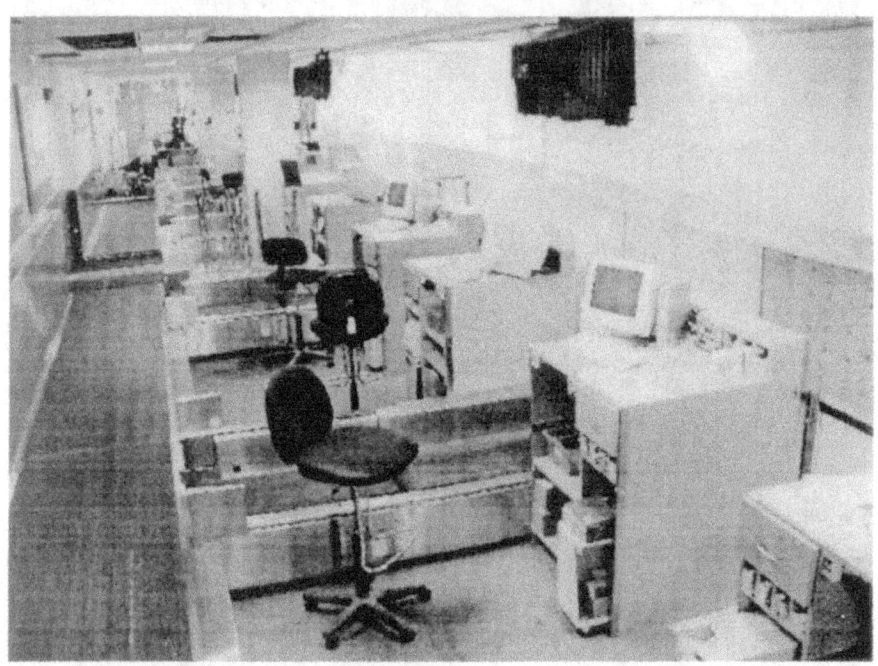

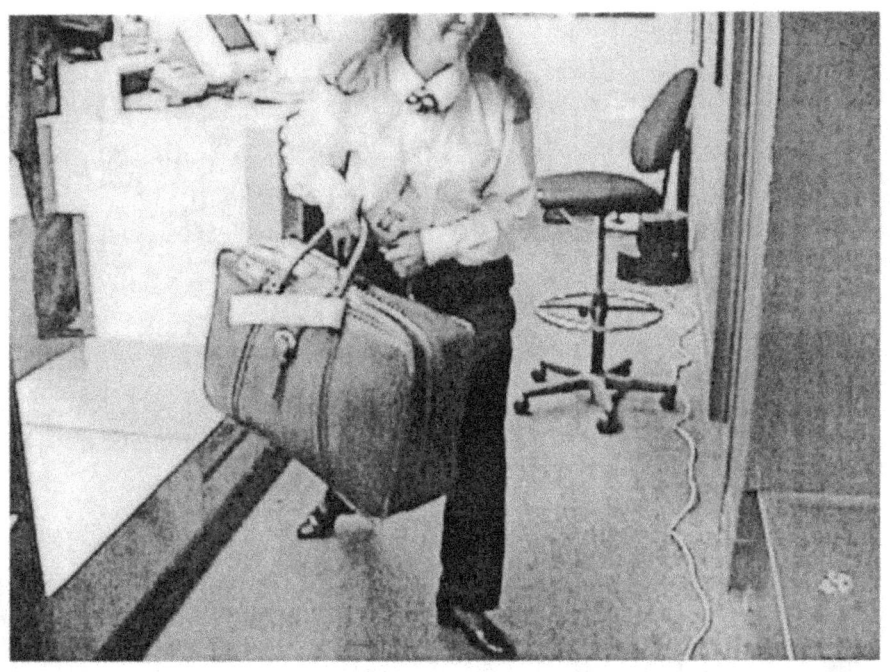

Index